Antenna Arrays and Automotive Applications

Contents

Part II Practical Design for Automotive

Abbreviations

ACC	Adaptive cruise control
A/D (ADC)	Analog to digital convertor
AM	Amplitude modulation
AOA	Angle of arrival
AMPS	Advanced mobile phone system uses 824–849 MHz to send information from the mobile station to the base station (uplink) and 869–894 MHz for forward channels (base to mobile)
AR	Axial ratio
BLIS	Blind spot information system
C2C (V2V)	Car to car (vehicle to vehicle)Automobile technology designed to allow automobiles to "talk" to each other
C2I	Car (vehicle) to infrastructure (communication)
CW	Continuous wave
DAB	Digital audio broadcasting
DBF	Digital beamforming
DBS	Direct broadcast satellite
DDC	Digital down converter
DF	Directional finding
DOA	Direction of arrival
DSRC	Dedicated short range communication
DVB	Digital video broadcasting
EIRP	Equivalent isotropic radiated power
ETC	Electronic toll collection
ESPAR	Electronically steerable passive array Radiator
FCC	Federal communications commission
FFT	Fast fourier transform
FM	Frequency modulation
FMCW	Frequency modulated continuous wave
GA	Genetic algorithm
GPS	Global position system
GSM	Global system for mobile communication

GSM 850	Global system for mobile communication system uses 824–849 MHz to send information from the mobile station to the base station (uplink) and 869–894 MHz for the other direction (downlink)
GSM 900	Global system for mobile communication uses 890–915 MHz to send information from the mobile station to the base station (uplink) and 935–960 MHz for the other direction (downlink)
GSM 1800	Global system for mobile communication uses 1710–1785 MHz to send information from the mobile station to the base station (uplink) and 1805–1880 MHz for the other direction (downlink)
GSM 1900	Global system for mobile communication uses 1850–1910 MHz to send information from the mobile station to the base station (uplink) and 1930–1990 MHz for the other direction (downlink)
HP	Horizontal polarization
HPWB	Half power beam width
IEEE	Institute of electrical and electronic engineers
IF	Intermediate frequency
IH	Interstate highway
ITS	Intelligent transportation system (service)
IVHS	Intelligent vehicle highway system
LHCP	Left hand circular polarization
LMS	Least mean square (algorithm)
LOS	Line of sight
LP	Linear polarization
LPF	Low pass filter
LRR	Long range radar
MIMO	Multiple input multiple output
MMIC	Monolithic microwave integrated circuit
MISO	Multiple input single output
MoM	Method of moments
MISIC	Multiple signal classification (algorithm)
MRC	Maximum ratio combining
MSE	Mean squared error
NATALIA	New automotive tracking antenna for low-cost innovative applications
NCO	Numerical controlled oscillator
NEC	Numerical electromagnetic code
NLOS	Non line of sight
OBU	On board unit
ORT	Open road tooling
PCB	Printed circuit board
PVC	Polyvinyl chloride
RCS	Radar cross section
RHCP	Right hand circular polarization
RF	Radio frequency
RFID	Radio frequency identification

RKE	Remote keyless entry
RMSE	Root mean square error
RSE	Remote start engine
RSSI	Received signal strength indicator
RSU	Road side unit
RTMS	Remote traffic microwave sensor
SDARS	Satellite digital audio radio service
SIMO	Single input multiple output
SISO	Single input single output
SLL	Side lobe level
SM	Spatial Multiplexing
SMILE	Spatial multiplexing of local elements
SNR	Signal to noise ratio
SP3T	Single pole, three throw
SP4T	Single pole, four throw (switch)
SPDT	Single pole, double throw (switch)
SPST	Single pole, single throw
STC	Space time coding
SRR	Short range radar
TPMS	Tire pressure monitoring system
UHF	Ultra high frequency
USS	Urban surface streets
VANET	Vehicular ad-hoc network
VHF	Very high frequency
VICS	Vehicle information and communication system
VP	Vertical polarization
VSWR	Voltage standing wave ratio
WAVE	Wireless access in vehicular environment
WLAN	Wireless local area network

Chapter 1
Introduction

1.1 Antenna Arrays in the Automotive Industry: Applications and Frequency Ranges

The evolution of wireless communication systems requires new technologies to support better quality communications, new services, and applications. In recent years, the automotive wireless communication market has expanded greatly. Modern cars use different services: AM/FM radio, satellite radio (SDARS), cellular phone communications, digital audio broadcasting (DAB), remote keyless entry (RKE), tire pressure monitoring (TPMS), and remote start engine (RSE) systems, TV, ubiquitous intelligent transportations systems (ITS), which include, global positioning system (GPS) data, electronic toll collection (ETC), vehicle information and communication service (VICS), collision safety radar devises, etc. [1].

It is well known that the antenna is the key element in achieving high performance of the communication system. Today, the directional antenna arrays or multiple antenna systems operating with diversity devices are becoming more and more popular for automotive applications. The advantages of the array antennas are very significant in the multipath fading environment caused by reflections from the surrounding buildings or from numerous cars on streets or highways. Over the past decade, research related to the applications of the antenna arrays in automotive wireless communications systems has exploded. The antenna arrays with the directional radiation pattern provide higher gain and immunity to the interference sources; smart antenna arrays dynamically adapt to the noise environment, producing nulls toward the noise sources and a high gain level toward the desired angle direction. Multiple-input multiple-output or MIMO systems, which use a few antenna elements from the transmitter or the receiver side (sometimes from both transmitter and receiver side), enhance the coverage area and improve multipath robustness.

V. Rabinovich and N. Alexandrov, *Antenna Arrays and Automotive Applications*,
DOI: 10.1007/978-1-4614-1074-4_1,

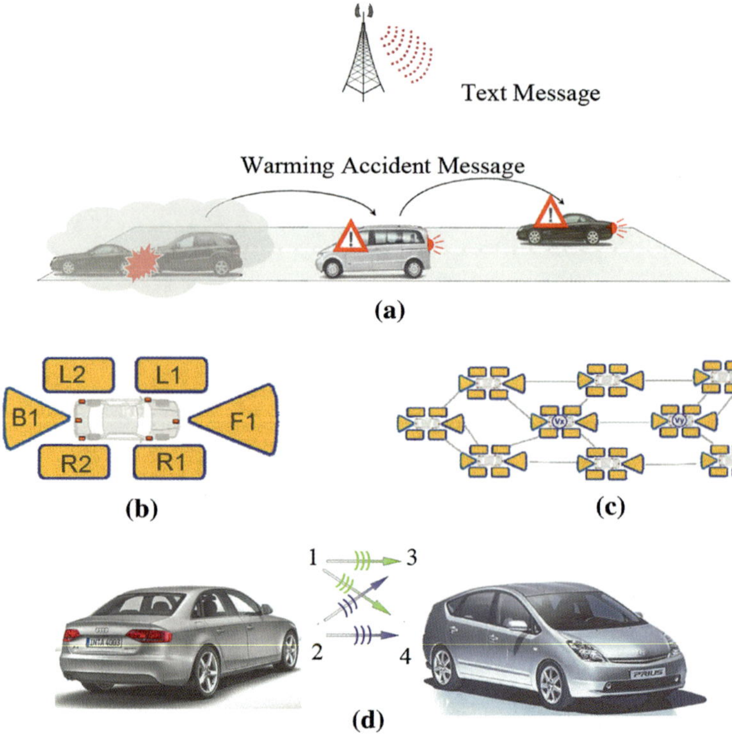

Fig. 1.1 Intelligent transportations systems concept: **a**, **b**, **c** Car to car and car to base station communication scenarios; **d** 2 × 2 MIMO car system topology

1.1.1 Intelligent Transportation Systems

1.1.1.1 Car to Car and Car to Infrastructure Communication Systems Including MIMO Design

It is well known [2] that intelligent transportation systems (ITS) containing transport infrastructure and vehicles improve transport safety, transport productivity, and travel reliability. One of the newest inter-vehicle and vehicle to infrastructure communication technology allows automobiles to "talk" to each other. This type of wireless system, which is based on the vehicular ad hoc networking (VANET) [3], allows cars to communicate with each other at distances approximately 100–300 m. The intelligent car to car (C2C) or car to infrastructure (C2I) communication, schematically shown in Fig. 1.1a, provides the driver with the following information:

- Road way conditions, warning on entering intersections and highways, reporting accidents and traffic jams, lane change warnings and low visibility ranges, collision safety distance between the cars

- Travel data about business locations, gas stations, and car services
- Specific speed limits, adaptable traffic lights, and other traffic management data
- Driver assistance including parking for vehicle, rest area location at the highway, finding the car on big parking lots, cruise control, etc.

The simplest link architecture utilizes the omnidirectional car-mounted antenna (covers all 360° around the vehicle), which supports local ad hoc wireless communication between its moving neighbors. More complicate multielement-directional car antenna systems improve the quality of reception in a multipath fading environment, extend communication range, and increase communication channel capacity. The architecture of high-throughput V2V networking with a directional steerable array is presented in Fig. 1.1b. The system consisting of six directional antenna arrays minimizes interference signals and scans maximum energy in six limited angle sectors named as: F1 as front; B1 as back; L1, L2 and R1 and R2 "look" left and right sides, respectively. Figure 1.1c shows the possible connectivity scenario between two neighboring vehicles moving at the same or different lines. Figure 1.1d shows the concept of a 2C2 MIMO system [4–7]: two mounted on one car omnidirectional antenna elements transmit signals (element numbers 1 and 2), and two antennas on the other car receive signals (element numbers 3 and 4). A simplified structure of the MIMO system is single-input multiple-output (SIMO) topology or multiple-input single-output (MISO) design. Each of the revealed configurations has its own features, strengths, and disadvantages. For example, diversity SIMO systems are relatively simple; however, the MIMO design offers tremendous potential for improving performances in a multipath environment.

The multiple element systems mounted on/in the car should be sorted into two main categories: compact-directional arrays with the interelement spacing less than the wavelength and the MIMO-distributed array elements with the spacing that can be more than the operating wavelength. The directional compact antennas concentrate energy between neighboring moving cars to share the communication information between them and minimize the interference signals coming from the other vehicles or surrounded objects. These antennas provide stable communication between the car and the base station.

Smart MIMO antennas increase the system capacity (bits/s/Hz), an important parameter that specifies the maximum data rate at which reliable communication between the transmitter and the receiver is possible. The capacity value depends on the number of transmitting and receiving antenna elements, as well as on the propagation channel statistic conditions between the transmitting and the receiving antennas. For example, for additive white Gaussian noise channel with the bandwidth B and the signal-to-noise ratio (SNR) γ, the Shannon capacity is given by

$$C = B \cdot \log_2(1 + \gamma) \tag{1.1}$$

If the communication system has N transmitting and N receiving antenna elements, and the transmitting power is equally divided between N antennas, the capacity value normalized to the bandwidth B becomes

$$C_N = N \cdot \log_2\left(1 + \frac{\gamma}{N}\right) \tag{1.2}$$

Formula (1.2) shows that when $\gamma/N \gg 1$, capacity is approximately proportional to the number of elements N.

Currently, numerous worldwide projects in Europe, USA (IEEE 802.11p standard in USA), and Japan [8–10] aim to develop electronic systems for inter-vehicle wireless communication. IEEE 802.11p is an approved amendment to the IEEE 802.11 standard to add wireless access in vehicular environments. It defines enhancements to 802.11 (the basis of products marketed as Wi-Fi) required to support ITS applications.

The standard IEEE 802.11p is used as groundwork for dedicated short-range frequency communication (DSRC), which is a short- to medium-range service developed to support vehicle-to-vehicle and vehicle-to-roadside communications. Such communication covers a wide range of applications, including vehicle-to-vehicle safety messages, traffic information, toll collection, drive-through payment, and several others.

VANET is currently in active development by General Motors, which demonstrates the system using Cadillac vehicles. Other automakers such as BMW, Daimler, Honda, Mercedes, and Volvo are also working on the C2C and the C2I communication systems. C2C and C2I communication techniques, described in a number of papers, reflect different design aspects of network configuration, such as common issues [11, 12], the antenna systems arrangement and design [13–15], processing algorithms based on smart antenna development [16–19], simulation, and measurement results [20–23].

The short-range communication network based on IEEE 802.11 technology can support Drive-Thru Internet service [24, 25]. Drive-thru Internet provides wireless local area network (WLAN) access for the moving vehicle during some (relatively short) periods of time by creating hot spots along the road—within a city or on a highway. One or more locally interconnected access points, as shown in Fig. 1.2a, form a so-called connectivity island that may provide local services as well as Internet access.

The WLAN operates in the following frequency bands: 2.4 GHz with the range of 2.4–2.485 GHz; 5 GHz with the ranges of 5.150–5.350 GHz and 5.470–5.725 GHz; and 5.725–5.850 GHz. Table 1.1 summarizes the main parameters of the various IEEE 802.11 standards. The data rate, shown in the table, are the maximum possible values and the range is the upper limit distance between communication devices.

Table 1.2 summarizes the main DSRC parameters.

Table 1.3 shows the frequency bands, which Japan uses for ITS.

The Federal Communications Commission (FCC) in the United States and the European Telecommunication Standard Institute (ETSI) also regulate electromagnetic radiation. For example, the effective isotropic radiated power (EIRP) defined by FCC Rules part 15.247 for an antenna with the gain of 6 dBi has to be less than 1 W.

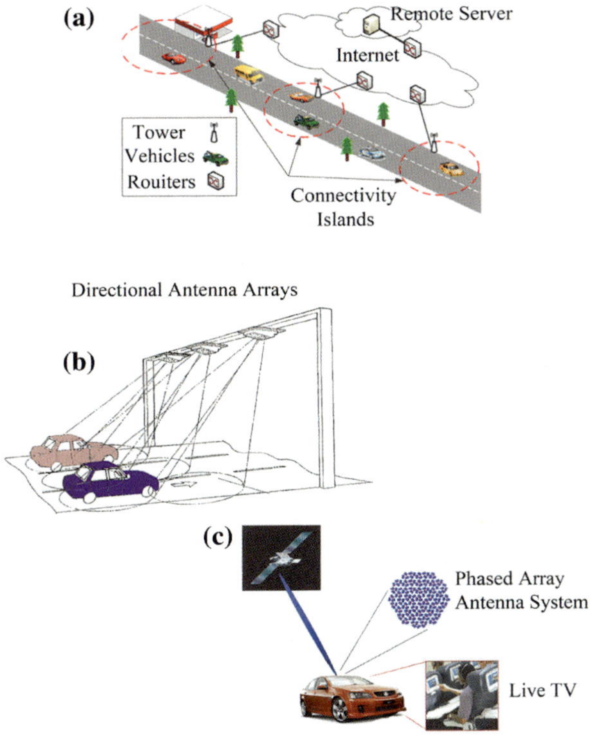

Fig. 1.2 **a** Drive-thru Internet concept; **b** Electronic toll collection topology; **c** Direct broadcasting satellite (DBS) TV communication

Table 1.1 IEEE 802.11 standard families

Standard	802.11a	802.11b	802.11g	802.11n
Frequencies (GHz)	5	2.4	2.4	2.4; 5
Range (max) (m)	70	100	100	200
Data rate (Mbps)	54	11	54	600

Table 1.2 DSRC technology: frequencies and communication characteristics

	902–928 MHz	5850–5925 MHz
Data rate (Mbps)	0.5	6–27
Maximum range (m)	100	1000
Channel capacity	1–2 channel	7 channel
Interference immunity	Low	High

The base station arrays consist of a dozen or more antenna elements, and there are no dimension requirements for these devices, which are, typically, mounted along the road. They have a medium gain G_{array}, typically, up to 15 dBi; the half beamwidth value varies from 30° to 60°, and they have dimension of about

Table 1.3 Frequency bands for Japan ITS

System	Service	Spectrum
VICS	Provide traffic information	76–90 MHz; 2.5, 5.8 GHz
ETC	Collect highway fee	5.8 GHz
DSRC	Provide various applications	5.8 GHz
Millimeter wave radars	Detect obstacles	24/26, 60/76, 79 GHz
Safety driving support system	Send safety information	700 MHz

Table 1.4 Antenna gain as a function of the half-power beamwidth for broadside direction

$\Delta\theta_x$, $\Delta\theta_y$ (degrees)	20	45	60	90	120
Gain (dBi)	19	12	9.5	6.0	3.5

0.5–2 m. These arrays are omnidirectional in azimuth and directional in the elevation plane, and they have a gain of about 6 dBi at 2.4 GHz and about 4 dBi at 5.8 GHz bands, respectively.

Vehicle-to-roadside or vehicle-to-vehicle antenna arrays do not have to degrade the esthetic car shape. This restriction dictates a compact car array design, sometimes mounted in the hidden parts of the car body, with two to four elements, and reduced interelement spacing. These antenna systems have a gain value less than 10 dBi and wider (compared to the base station arrays) beamwidth. Simple formulas allow estimating the array gain as a function of the antenna beamwidth in two orthogonal planes $\Delta\theta_x$ and $\Delta\theta_y$(in degrees) [26]

$$G_{planar} \approx \frac{32400}{\Delta\theta_x \cdot \Delta\theta_y} \cdot \eta \tag{1.3}$$

Coefficient η determines the loss factor of the feeding network and array elements.

Table 1.4 presents gain values calculated according to the ratio (1.1) assuming that $\eta = 1$ (lossless antenna).

1.1.1.2 Electronic Toll Collection Design

The ETC system [27], operating in DSRC spectrum, automatically collects tolls at toll roads. ETC is created as an alternative to manual payment collection in toll plazas. One of the earliest implementations of ETC in the U.S. was on the Dallas North Toll way in 1989. Since then, numerous sites around the world have implemented ETC to increase roadway revenues and easing traffic congestion. The automatic ETC system based on RFID technology is composed of an on-board unit (OBU) with tag antenna, a road side unit (RSU) with the reader antenna, and the other components. When the vehicle passes the toll station, RSU sends out the RF question signal, then the OBU responds and the center computer control system identifies vehicle plate number, user name, mailing address, driver's license, and the toll is deducted from the user account. ETC can be implemented in different ways. The simplest approach makes use of a toll plaza with individual lanes that separated by barriers. Such method reduces the flow rate, but prevents vehicles from passing through if thc toll has not been paid. An alternative approach is an open road tolling (ORT). It uses no barriers, no toll booths. An advanced system includes an overhead reader and a transponder mounted on/in the vehicle. When a vehicle equipped with a transponder passes through a capture zone, the vehicle location is associated with the specific travel lane at that instant in time, and the short length of the zone allows for accurate timing alignment with the vehicle detection systems. The major advantage of ORT is that users are able to drive through the toll plaza at highway speeds without having to slow down to pay the toll. A communication scenario of the toll collection system with a directional antenna is shown in Fig. 1.2b. The signal is transmitted by a roadside antenna array and received by the omnidirectional antenna (tag antenna) mounted on the front car windshield. The transponder responds by generating a microwave signal, which is coded by data stored in the tag. The transmitted signal is received by the roadside antenna array system, which decodes the data and relays this information to a computer to identify the vehicle. ETC base station antenna array provides fast and reliable car detection at the entrance and exit of highway lines. ETC array can operate as a fixed beam system, which covers only one vehicle lane, or as electronic phased array that covers a few driving lanes. Let us introduce a concept, "action range", that is the region on a road illuminated by the antenna, in which the RFID Tag can interact with the reader antenna. It should be designed to a rectangular region as wide as the roadway lane, should have enough length L, which is determined by the velocity of the car, and the operation time of ETC system. According to the basic requirements for global ETC published by Japan highway industry development organization [27] the length L of active communication zone for antenna covering a single lane is about 4 m when the reader antenna is mounted at 5–6 m above the highway. It is necessary to note that the active zone should not be too long. If there are two or more cars in one action range simultaneously, confusion can easily occur.

Several reader antenna parameters are considered as key antenna characteristics:

1) Sufficient bandwidth to get enough data transferring velocity.
2) Sufficient antenna gain to obtain read-and-write distance between the Reader and the Tag.
3) Proper main lobe width to cover required capture zone and low side-lobe level.
4) The signal polarization of the reader antenna is an important factor. For example, the horizontally polarized signal experiences greater attenuation through the windshield, where a tag transponder with an antenna is installed. Also, data indicate that horizontally polarized signals may lead to a higher level of fluctuations in the multipath reflection environment. However, the car body can change the polarization of the signal transmitted by the Tag system.

Therefore, vertical, dual, or circularly polarized reader antenna designs are acceptable. A typical beamwidth of the ETC antenna array is about 30° in the horizontal plane and more than 30° in the vertical plane.

Frequency ranges for ETC are as follows: 866–869 MHz in Europe, 902–928 MHz in North and South America, and 950–956 MHz in Japan and some Asian countries. The new standard uses frequencies between 5.850 and 5.925 GHz of DSRC band in North America, and 5.8 GHz in Europe. The new frequency range provides 75 MHz of frequency bandwidth (915 MHz frequency has only 12 MHz of frequency bandwidth) and permits much higher data transmission rates (6–27 Mbps) than the lower frequency 915 MHz band (0.5 Mbps) and channel capacity up to 7 in comparison with 1–2 for a 915 MHz band.

1.1.1.3 Radar Technology for Automotive Applications

Automotive radars are currently very attractive, although radar technology has been investigated for use in vehicles since the 1970s. There are two different types of automotive radar systems with antenna arrays commercially available: Long-range radar (LRR) and Short-range radar (SRR) [28–30]. Figure 1.3a illustrates the variety of applications of the LRR and SRR systems. The LRR operates at 77 GHz and has a resolution of about 3 m at a distance of 150 m.

The 77 GHz radar technology is used in adaptive cruise control (ACC) applications. One or a few LRR's main lobes control the driving path in front of the car to determine the constant minimum safety distance relative to the vehicle in front.

Typical performance specifications of an ACC system with antenna are as follows:

a) Transmit frequency = 76–77 GHz;
b) The mean transmitted power level <50 dBm (peak level is 55 dBm);
c) Target range = 2–150 m;
d) Range resolution ±1 m;
e) Velocity resolution ±1 km/h,
f) Azimuth Angular Coverage = ±8° with 3° minimum resolution;
g) Elevation Angular Coverage = 3°–4° (single beam); and
h) Antenna gain = 26–34 dBi.

The SRR system operates at 24 GHz with a range less than 30 m. The SRR system is used to provide side, rear, and front crash safety, intersection safety, blind spot safety, and other kinds of safety enhancements.

Antenna specifications for SRR devices include:

a) Transmit frequency = 22–26 GHz;
b) The mean transmitted power level <0 dBm (peak limit is 20 dBm);
c) Target detection range = 0.05–25 m;
d) Azimuth Angular Coverage = 55° (typical, 3 dB beamwidth);

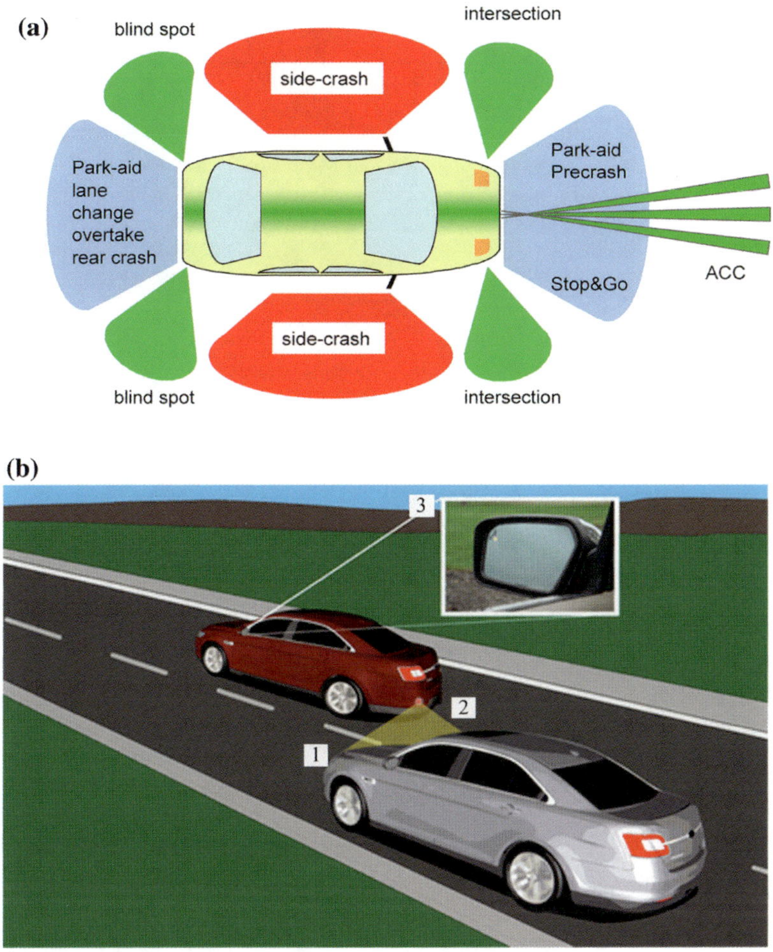

Fig. 1.3 **a** Long- and short-range radar applications; **b** Blind spot information system concept (1-Passing vehicle enters blind spot of the lead vehicle; 2-Lead vehicle's radar identifies passing vehicle in blind spot; 3-Indicator light illuminated on corresponding side mirror)

Table 1.5 Summary of key parameters for SRR system

	Blind spot	Parking aid	Stop and go	Simple pre-crash
Maximum detection range (m)	4–8	2–5	20	7–10
Required range resolution (m)	0.1–0.2	0.05–0.2	0.2–0.5	0.1–0.2
Minimum object size	Bicycle	3″ PVC Pole	Bicycle	Metal post

e) Elevation 3 dB beamwidth = 15° typical,
f) Range resolution 0.05–0.2 m (depends on the application); and
g) Range Accuracy = ±5 cm.

Table 1.6 LRR and SRR frequency bands

Applications	Safety aspect	Technology
Adaptive cruise control	Normal driving; accident avoidance	77 GHz radar
Pre-crash	Accident; mitigation of impact	77 GHz/24 GHz radar; 76 GHz/81 GHz radar
Blind spot detection	Normal driving; accident avoidance	24 GHz Radar/Vision Sensor
Lane departure warning	Normal driving; accident avoidance	Vision Sensor
Stop and go	Normal driving; accident avoidance	77 GHz/24 GHz radar; 76 GHz/81 GHz radar

Frequency bandwidth of the SRR is about of 4–5 GHz.

Table 1.5 shows a summary of some key parameters for the SRR system.

Table 1.6 sorts LRR and SRR according to frequency bands.

A lot of automotive manufacturers utilize automotive radar devices: Daimler–Benz, BMW, Jaguar, Nissan, Toyota, Honda, Volvo, Ford, GM, etc.

The blind spot information system (BLIS) (concept is shown in Fig. 1.3b) was recently introduced by Ford in a few car models. The system helps to detect vehicles in blind spots during normal driving and traffic approaching from the sides when reversing out of parking spots. BLIS uses two multibeam radar modules, which are packaged in the rear quarter panels—one per side. The radar recognizes when a car enters the blind spot zone, and an indicator on the corresponding side view mirror provides a warning signal that a vehicle is approaching.

Delphi Rear and Side Detection System (RSDS) developed a single beam monopulse radar, which helps to make drivers aware of approaching vehicles when changing lanes or making turns. Delphi's Collision Mitigation System (CMC), based on the radar technology and vision sensors, warns of a pending collision when the vehicle approaches a pedestrian or another vehicle. Promising results were obtained for 24 GHz SRR radar operating with the Digital Beamforming processor [30]. This system uses high resolution MUSIC algorithm which significantly improves regular results corresponding to the Rayleigh criterion.

1.1.2 Space Diversity System for FM, Digital Terrestrial TV, and RKE/TPMS Devices

Antennas with two or three elements are widely used for FM broadcasting radio diversity systems [1], which cover the frequency range from 65 to 108 MHz. North America uses the band from 87.5 to 108 MHz, Japan radio stations operate within 76–90 MHz, Eastern Europe from 65.8 to 74.0 MHz, although a number of these countries have added the 88–108 MHz band and are using the two FM bands simultaneously, as in the case of Russia. Space diversity is a simplified MIMO technique named as SIMO method. SIMO systems are good candidates for use in the terrestrial digital video broadcasting (DVB) TV devices mounted on cars. DVB covers the VHB band from 110 MHz to 270 MHz and UHB band from 350 MHz to 870 MHz (depending on the country).

It is well known that SIMO (1 × 2) schemes based on two receiving antennas significantly increase the communication range of RKE/RSE system and improve the quality of the TPS devices. Figure 1.4 demonstrates the conceptual logic of the SIMO system operation in the condition of the multipath fading propagation when the signal received by the antenna system fluctuates due to the multiple reflections from the building and moving cars. Figure 1.4a schematically demonstrates multiple propagation scenario between the radio station and the car, Fig. 1.4b shows a block diagram of the simplest selection diversity system with two antennas, two low-noise amplifiers, an electronically controlled switch, and a control logic circuit. According to this design, the output of the receiver detects the RF signal, and the control logic circuit (according to the specific algorithm) connects either the first or second antenna to the receiver. The simplest algorithm includes the signal estimation, and when the signal level received by the operating (first) antenna falls below the predetermined level, the switch connects the second antenna to the car radio. A more complicated algorithm includes the estimation of four signal levels: from the first antenna, from the second antenna, from the first and second antennas together in phase, and from the first and second antennas out of phase 180°. Figure 1.4c shows the radiation patterns of an RKE antenna located in two different car body places, and Fig. 1.4d demonstrates the combined "diversity radiation pattern". Measurements show that, when using a single antenna, the communication range varies from almost 100 m (toward the angle corresponding radiation pattern maximum) to 10 m (toward the angle corresponding radiation pattern minimum). When using diversity system, the variations are only 20 m from the maximum communication range.

1.1.3 Direct Broadcasting Satellite TV

Direct Broadcasting Satellite (DBS) TV steerable beam antenna arrays [31–34] mounted on the car roof (Fig. 1.2c) operate in microwave Ku-band frequency range from 12.2 to 12.7 GHz in North and South America, 10.7–12.75 GHz in Europe, Russia, and Africa), and from 11.7 to 12.2 GHz in Asia and Australia. Passengers of a car can watch satellite television programs while their vehicle is stationary or in motion. Therefore, the system needs to be equipped with a special sensor that provides minimal searching to find the satellite. The tracking process has to keep the maximum radiation antenna pattern at the angle position that corresponds to the satellite angle direction. The main component of the antenna system is a phased array. Classic phased arrays consist of hundreds or even thousands of tiny antenna elements. Relative phases of the different elements vary in such a way that the antenna array radiates maximum energy in a predetermined angle direction. Satellite TV arrays are designed to receive circularly polarized or

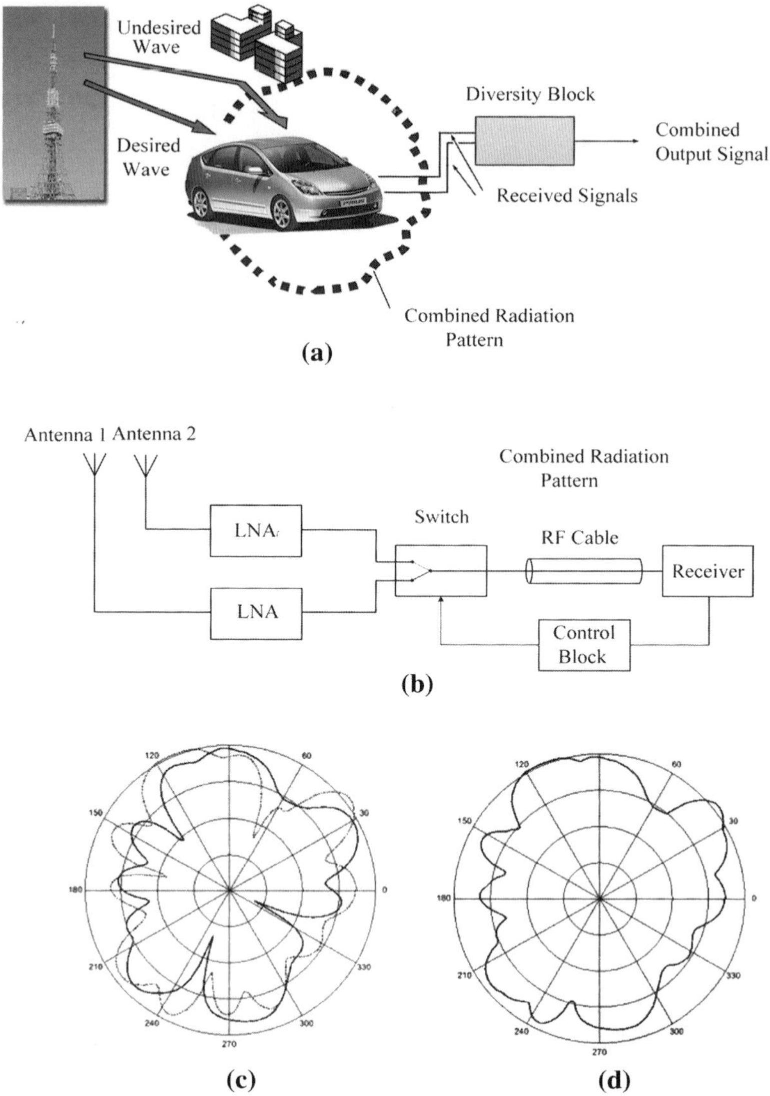

Fig. 1.4 **a** Multiple propagation scenario for car-base station communication; **b** Simple logic for diversity reception with two antennas; **c** Radiation pattern example for individual car antenna; **d** Diversity radiation pattern

linearly polarized waves with electrical adjustment of polarization plane waves. For example, arrays designed for North and South America can pick up signals with both right-hand (RHCP) and left-hand (LHCP) polarization. The antenna beamwidth varies from 1.6 to 3.8° in both the elevation and azimuth directions with the antenna gain of 28–34 dBic. The antenna has to provide 360° azimuth scanning, and the elevation scanning angle range depends on the TV satellite location. For example, complete coverage over the continental US requires an elevation angle range of 20°–65° with Direct TV and 15°–60° for EchoStar. The tracking accuracy has to be about 0.5° and 2.0° in elevation and azimuth, respectively. The polarization is switchable between right-hand circular polarization (RHCP) and left-hand circular polarization (LHCP). Cross polarization isolation is greater than 25 dB. Today, low cost microstrip antenna array is an excellent candidate for the automotive satellite TV applications.

1.2 Antenna Arrays: From Simple Configuration to Advanced Design

The arrays can be categorized as antennas with fixed directional beam or antennas with electrically (or mechanically) steered beam. A system with fixed beam pointing is the simplest configuration of the directional antenna that provides higher (in comparison with omnidirectional design) gain and lower sidelobe level. A more complicated structure is the directional antenna which provides narrow steering beam within the required angle sector and adaptive nulling toward the noise source directions. In noise environment, these arrays are adapted to the dynamically changing traffic volume, communication congestion, or degradation of radio wave propagation characteristics. Such antenna array design, of course, is not simple as the fixed beam array because it includes a special electronic control system to provide the required beam steering.

1.2.1 Fixed Beam Directional Arrays

Today, different fixed beam antenna arrays are used for wireless connectivity between the base station and the electronic device mounted on/in a moving car. Typical concepts of fixed beam directional microstrip arrays are shown in Fig. 1.5a–e. Figure 1.5a presents linear array with corporate (parallel) beamforming, Fig. 1.5b and Fig. 1.5c shows planar 2×2 and 4×4 arrays with a parallel feeding network. Figure 1.5d demonstrates a linear array with series feeding network, and Fig. 1.5e relates to the planar array with combined series–parallel network design. It is necessary to note that combining the single elements into the array requires proper matching to provide minimum reflections at the array output. For example, two single patches, each with 50 ohm output impedance, can be combined if each of them is connected with a quarter wave microstrip transformer with an impedance equal to 71 ohm. Such transformer line converts 50 ohm into 100 ohm. So, when two elements with output impedances equal to 100 ohm form two element array with the parallel feeding network, antenna output impedance becomes 50 ohm. Impedance of the microstrip line is determined by the line width. For example, a microstrip line printed on the FR4 dielectric material (dielectric

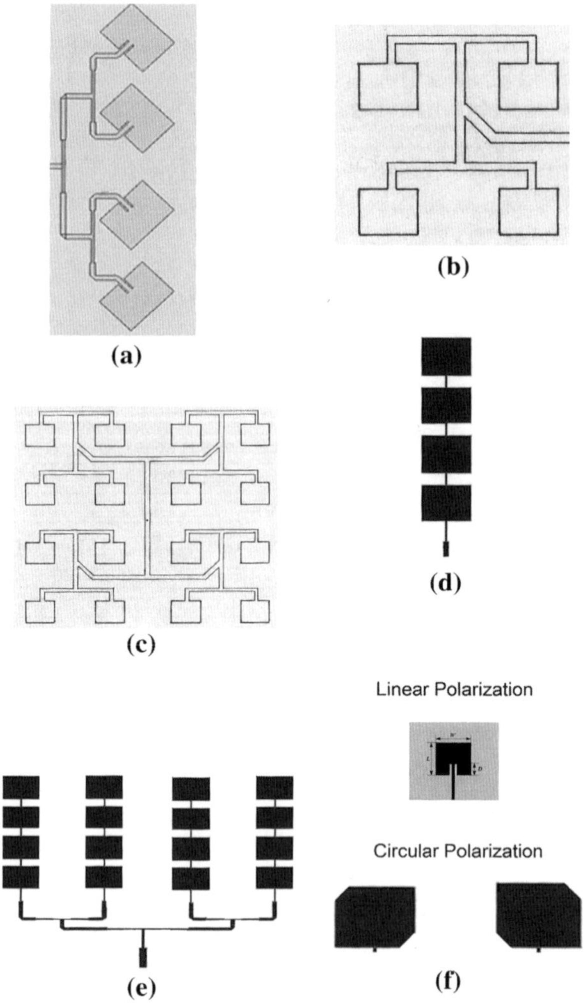

Fig. 1.5 Topologies of fixed beam directional antenna array: **a**, **b**, **c** Parallel feeding; **d** Series feeding; **e** Combined parallel/series feeding; **f** Patch antenna element design for linear and circular polarization

constant is 4.6 and thickness is 1.6 mm) has width w = 3 mm for 50 ohm impedance, w = 1.57 mm for 71 ohm impedance, and w = 0.65 mm for 100 ohm impedance. Figure 1.5f shows examples of patch elements forming different types of polarization: linear and circular.

1.2.2 Arrays with Simple (ON/OF) Electronically Switchable Elements

A typical example of ON/OFF electronically switchable steerable array is shown in Fig. 1.6a. The antenna consists of one RF-exited center element [35–37] and a few circularly mounted parasitic elements. Each parasitic element is connected to

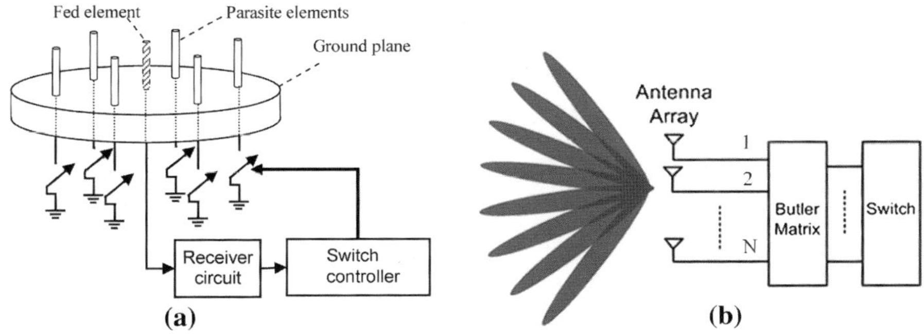

Fig. 1.6 **a** Example of ON/OFF electronically switchable array; **b** Concept of the array with Butler matrix

a ground plane via a switch that has two states: "ON" when a corresponding parasitic element is connected to the ground or "OFF" when element is not active (disconnected from the ground). Reconfiguration of the radiation pattern shape is provided by connecting a few parasitic elements to the ground using pin-diode switches with a switching time of a few nanoseconds.

An antenna array with Butler matrix provides steering of the radiation pattern. The Butler matrix is a type of the beamforming network. The linear array with the Butler matrix feeding network presented in Fig. 1.6b generates either a multibeam design if each output array port is connected with its own receiver/transmitter or a steering system with a single receiver/transmitter and an electronically controllable switch that sequentially connects each output port with a single receiver/transmitter. Depending on which of the N inputs is accessed, the antenna beam is steered in a specific direction in one plane. The popularity of the Butler matrix operating equally well in transmit and receive mode is due to the easy implementation of hybrids and fixed phase shifters. Arrays with the Butler Matrix circuit and ON/OFF switchable systems are more complicated than the fixed arrays, but still easier to implement than phased arrays with electronically controlled phase shifters and sophisticated adaptive arrays.

1.2.3 Electronically Controlled Phased Antenna Arrays with Single Output Receiver

These devices [38] are widely used in the defence industry, however, they are becoming more popular in civil and commercial applications. Beamforming networks with the electronically controllable elements are shown in Fig. 1.7a–d. Figure 1.7a presents popular block diagram of feeding network with electronically controlled phased shifters. The phase shifter of each antenna array element has a special electrical control circuit (for example, pin-diode circuit), which can change the phase of the received signal. According to the predetermined steering angle θ_0 the computer calculates phases for each array element to direct the beam (maximum receiving) at

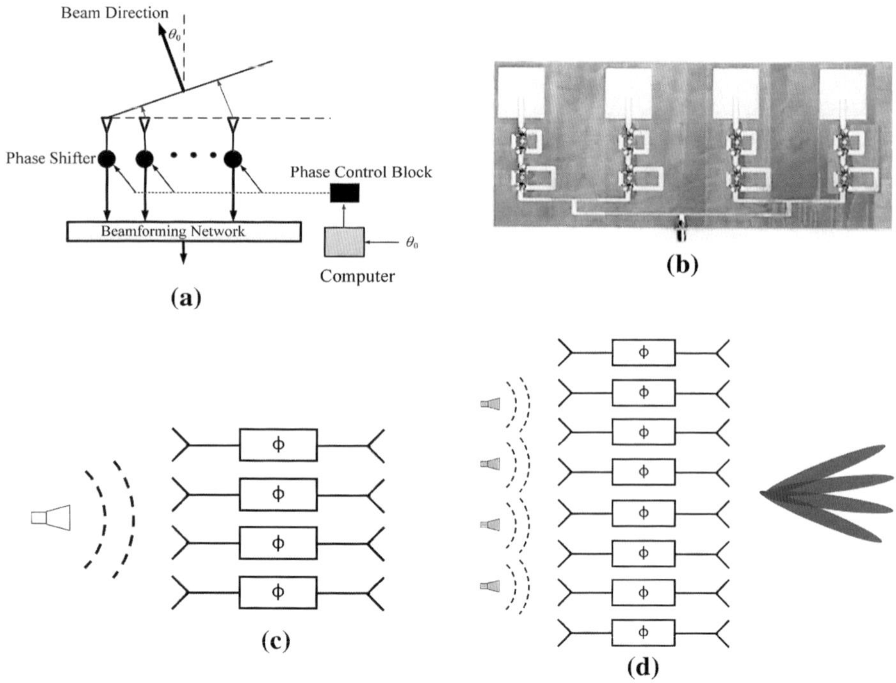

Fig. 1.7 Beamforming networks for electronically controllable array: **a** Block diagram of the array with phase shifters; **b** Four-element array with patch elements and pin diode phase shifters; **c** Space fed array; **d** Multibeam space fed array

a certain angle position. The phase control block converts the calculated phases into the electrical signals, and these control signals change the phase shifters "states," collimating maximum energy at the given angle direction. Figure 1.7b demonstrates prototype example of four element array with two-bit digital pin-diode phase shifters. Typically, phase shifters used for phased array are built as digital phase shifters. It means that practical phase shifter can realize only a discrete number of phase states, for example, a two-bit shifter realizes four phase states with a discrete equal to 90° within 360°. Such systems are not very complicated when using a small number of the array elements and two- or three- (not more) bit phase shifters.

The space fed array shown in Fig. 1.7c consists of feed antenna, for example, horn, array elements, phase control block, and computer. In the transmitting mode, feed horn illuminates array elements located at the feed side. Illuminated waves pass through the antenna element channels, undergo a phase shift, and radiate into space by the antenna elements represented by the v-shaped symbols on the right of the array (Fig.1.7d). The reverse occurs for the energy reception: the elements on the right of the array capture the energy from the source, the phase shifters shift the wave phases, and the antenna elements on the left of the array focus and radiate the energy to the feed. Phase shifters convert spherical illuminated wave into the plane wave and

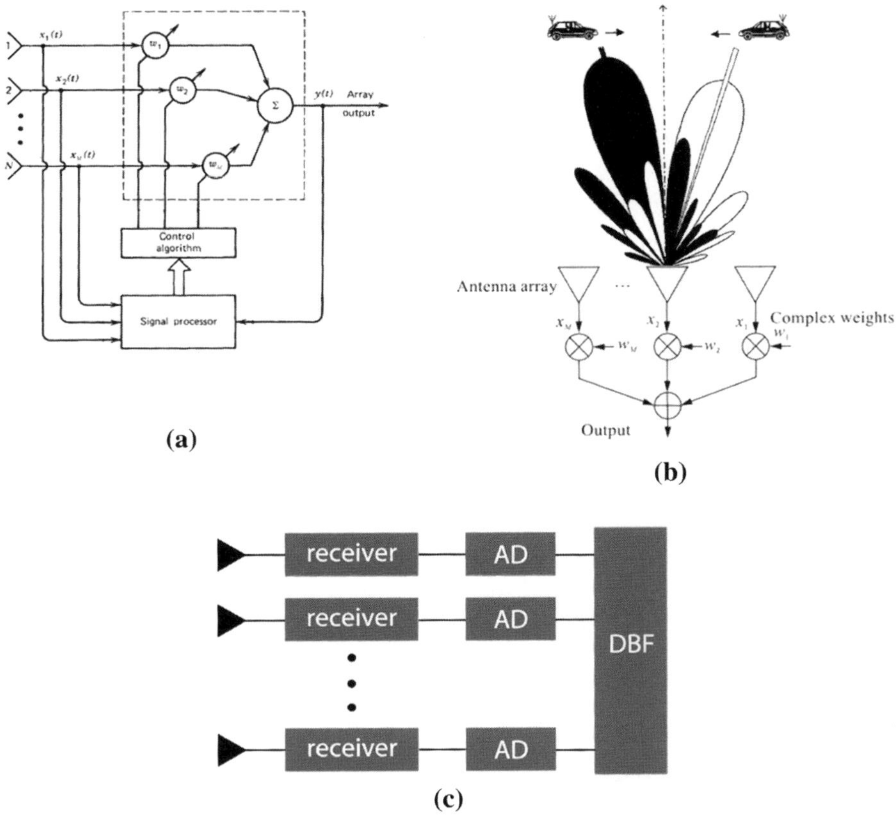

Fig. 1.8 Adaptive array block diagrams: **a** analog adaptive weighting; **b** Concept of communication with moving cars; **c** Digital beamforming (digital processor produces adaptive control)

provide main beam steering as indicated in the previous discussion. Generally, the feed gain pattern (from the feed antenna) is adjusted to achieve a desired sidelobe level for the overall antenna system. In order to obtain a good tradeoff between gain and sidelobe levels for the overall antenna, the feed pattern of the horn has to provide the amplitude level at the edge of the array between 10 and 20 dB below the peak value. Such a design does not need microstrip feeding lines and, therefore, provides reduced losses in comparison with the other systems. A similar array with a few feedings elements shown in Fig. 1.7d, provides multibeam mode operation or switch regime if the single receiver is connected with one of the feeding elements.

1.2.4 Adaptive Antenna Arrays

Adaptive antennas are the smartest antenna systems. Each antenna element of the array [39, 40] is associated with a weight (Fig. 1.8a) that is adaptively updated so that the array gain in a particular look direction is maximized, while toward interfering signals is minimized. In other words, weight coefficients change the reception pattern dynamically to adjust to variations in channel noise and interference in order to improve the SNR of a desired signal. An intuitive operation of the adaptive antenna system can be obtained from Fig. 1.8b, where the weight coefficients W_m (m = 1, 2…M; M = the total array element number) are adapted to receive the desired user signal and to suppress the noise source. The dark beam pattern is used to communicate with the user on the left, while the light beam is used to communicate with user on the right. It is seen that each pattern has nulls in the direction of the other user. As the user moves, the beam patterns are continuously updated, adapted to the maximum signals toward the desired sources, forming nulls in the directions of the interference signals. Unlike the switched beam approach, the adaptive system is able to continuously distinguish between the signal and the interference, while maintaining enhanced carrier to interference ratio levels. Conventional mobile systems usually employ some sort of antenna diversity (e.g., space, polarization, or angle diversity). These antennas provide the best performances in multipath propagation conditions; however, they are complicated structures, and demand special hardware to regulate amplitude and phase at each array element. Adaptive systems can be designed as analog devices or with a digital beamforming processor, as shown in Fig. 1.8c. Each antenna element has a receiver and analog to digital convertor (ADC). Digitized signals are processed in a digital beamformer (DBF) to create a digital radiation pattern that provides required parameters. A digital signal processing enables adaptive multi-user beamforming, a type of processing that has no equivalent in the analog radio frequency (RF) domain. As a result, in multiple access systems, smart antennas may be combined with interference-canceling multi-user receivers. A digital signal processor that uses high resolution techniques significantly improves angle resolution when estimating the angle directions of the incoming signal sources.

References

1. Rabinovich V et al (2010) Automotive antenna design and applications. CRC Press, Boca Raton
2. Huang X (2006) Smart antennas for intelligent transportation systems. 2006th international conference on ITS telecommunication proceedings, pp 426–429
3. Hartenstein H, Laberteaux K (2010) Vehicular applications and inter-networking technologies (VANET). Wiley, New York
4. Karedal J (2009) Measurement-based modeling of vehicle-to-vehicle MIMO channels. IEEE ICC 2009 proceedings, Dresden, Germany
5. Laurenson et al D (2009) Modeling for vehicle to vehicle applications. Loughborough antennas and propagation conference, pp 42–48

6. Matthaiou M et al (2008) Capacity study of vehicle-to-roadside MIMO channels with a line-of-sight component. Wireless communication and networking conference, IEEE, pp 775–779
7. Reichardt L et al (2009) Performance evaluation of SISO, SIMO and MIMO antenna systems for car-to-car communications in urbanon environments. Intelligent transport systems telecommunications, 9th conference. pp 51–56
8. www.car-to-car.org
9. Bell MGH (2006) Policy issues for the future intelligent road transport infrastructure. IEE proceedings on intelligent transport systems 153(2):147–155
10. http://www.nhtsa.gov
11. Wu S, Tseng Y (2007) Wireless ad hoc networking: personal-area, local-area, and the sensory-area networks, Auerbach, New York
12. Misra S et al (2009) Guide to wireless ad hoc network. Springer, New York
13. Kornek D et al (2010) Effects of antenna characteristics and placements on a vehicle to vehicle channel scenario. ICC 2010 workshop on vehicular connectivity, May 2010, Cape Town, South Africa
14. Thiel A et al (2010) In situ vehicular antenna integration and design aspects for vehicle-to-vehicle communications. Antennas and propagation EUCAP, proceedings of the 4 European conference
15. Pontes J et al (2011) Investigation on antenna systems for car-to-car communication. IEEE J Sel Areas Commun 29(1):7–14
16. Lee S et al (2007) Performance analysis of beamforming techniques in ad-hoc communication between moving vehicles. Proceedings of Asia-Pacific conference on communications, pp 185–188
17. Yamamoto K et al (2010) Improving performance of inter-vehicle communication using LMS adaptive circular array antenna. 2010 IEEE intelligent vehicles symposium university of california, San Diego, CA, USA June 21–24
18. Mase K (2008) Performance evaluation of a roadside-to-vehicle communication system using narrow antenna beam switching based on traffic flow model. Globecom workshops, 2008 IEEE
19. Vu V et al (2006) Digital and super-resolution ultra wide band inter-vehicle localization system. Communication and electronics (ICCE), first international conference, pp 446–450
20. Brown T et al (2011) Artificial wideband multi user channels for rural high speed vehicle to vehicle links. IEEE J Sel Areas Commun 29(1):29–36
21. Acosta G, Ann Ingram M (2006) model development for the wideband expressway vehicle-to-vehicle 2.4 GHz channel. Wireless communications and networking conference IEEE, pp 1283–1288
22. Ramanthan R et al (2005) Ad hoc networking with directional antennas: a complete system solution. IEEE J Sel Areas Commun 23(3):496–506
23. Subramanian A et al (2008) A measurement study of inter-vehicular communication using steerable beam directional antenna. VANET'08, 15 September, 2008, San Francisco, California, USA
24. Ott J (2004) The Drive-thru architecture: WLAN based internet access on the road. Vehicular technology conference, IEEE, pp 2615–2622
25. Tan W et al (2011) Analytical models and performance evaluation of drive-thru internet systems. IEEE J Sel Areas Commun 29(1):207–222
26. Balanis C (1997) Antenna theory, analysis and design, 2nd edn. Wiley, New York
27. http://www.hido.or.jp/ITSHP_e/
28. Schneider M (2005) Automotive radar—status and trends. German microwave conference, pp 144–147
29. Gresham I et al (2004) Ultra-wideband radar sensors for short-range vehicular applications. IEEE Trans Microw Theory Tech 52(9):2105–2122
30. Wenig P et al. (2007) System design of a 77 GHz automotive radar sensor with superresolution DOA estimation. Signal, systems and electronics, international symposium, pp 537–540

31. Wang J, Winters J (2004) An embedded antenna for mobile DBS. Vehicular technology conference, 60th IEEE, pp 4092–4095
32. Hule F et al (2005) A direct broadcast reception system for automotive OEMs. IEEE antennas and propagation society symposium, 2005 IEEE, pp 80–83
33. Wang J et al (2008) Vehicle mounted satellite antenna embedded within moon roof or sunroof, US Patent 7,391,381, June 2008
34. Vaccaro S et al (2011) Low cost ku-band electronic steerable array antenna for mobile satellite communications. Antennas and propagation (EUCAP), Proceedings of the 5th European conference on 2011, pp 2362–2366
35. Donelli M et al (2007) A planar electronically reconfigurable wi-fi band antenna based on a parasitic microstrip structure. IEEE Antennas Wirel Propag Lett 6:623–626
36. Sibille A et al (1997) Circular switched monopole arrays for beam steering wireless communications. Electron Lett 33(7):551–552
37. Thiel D (2004) Switched parasitic and controlled reactance parasitic antennas: a systems comparison. Antennas and propagation society international symposium, 2004 IEEE, pp 3211–3214
38. Hansen R (2009) Phased array antennas. Wiley, New York
39. Monzingo R et al (2009) Introduction to adaptive arrays, 2nd edn. SciTech, Perth
40. Primak S, Kantorovich V (2011) Wireless multi-antenna channels: modeling and simulation. Wiley, New York

Part I
Performances and Processing Methods

Chapter 2
Typical Array Geometries and Basic Beam Steering Methods

2.1 Introduction

Any electronic system with an antenna array consists of two or more antenna elements, beam forming network, and a receiver or transmitter. Many different antenna configurations can be utilized as an antenna element in the antenna array: simple dipole, monopole, printed patch design, Yagi antenna, etc. The primary requirement for antennas mounted on/in the car is compact and aesthetic design.

Several locations are considered to use for mounting of the antenna elements on/in the vehicle [1–7, 12]: roof, car window glasses, bumper, or trunk, front panel, side, and rear view mirrors, etc. A typical antenna element mounted on the car roof is a wire or printed on the circuit board monopole. Car glass printed strip antenna elements are proposed for receiving AM/FM radio, TV, and RKE/RSA radio signals, as well as for ITS communication between mowing cars or between the vehicle and roadside base station systems [8−11]. Anti-collision antenna arrays [7] are mounted on the car bumper. Typically, antenna array elements are configured as an equally spaced linear design, planar geometry with equal spacing between adjacent elements in each column and row, or antenna elements arranged in a circle with an equal spacing between adjacent antennas. SIMO diversity systems [4, 6] do not require regular arrays with the predetermined spacing between the antenna elements and can use distributed elements with an inter-element spacing more than operating wavelength. Generally, inter-element spacing has to be chosen so to ensure low-level correlation between signals received by adjacent antenna element channels. According to the theoretical results, the low correlation coefficient is obtained for inter-element spacing more than half wavelength; however, experimental measurements show that this distance can be less [13]. Therefore, both compact antenna arrays and distributed multi-element antenna systems are good candidates for automotive car applications.

V. Rabinovich and N. Alexandrov, *Antenna Arrays and Automotive Applications*,
DOI: 10.1007/978-1-4614-1074-4_2,

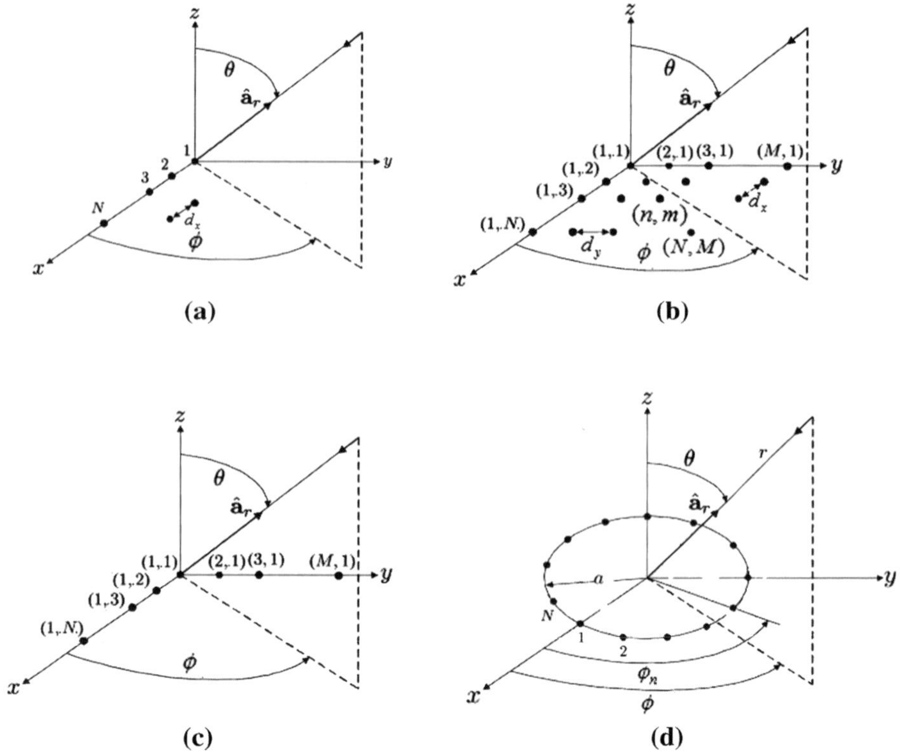

Fig. 2.1 Spherical coordinate system: **a** Linear array; **b** Planar array; **c** Mills cross array; **d** Circular array

2.2 Linear Array Factor for Equally Spaced Elements

The array factor of linear array with equally spaced N isotropic radiating elements placed along horizontal axis x as a function of the angles θ and ϕ in a spherical coordinate system shown in Fig. 2.1a can be expressed by simple formula

$$AF_{\text{linear}}(\theta, \phi) = \sum_{n=1}^{N} I_n \cdot e^{j\cdot(\delta_n + k\cdot d\cdot n\cdot \sin\theta)} \qquad (2.1)$$

where I_n and $\delta_n(n = 1, 2 \ldots N)$ are the amplitude and phase excitation of nth array element, $d =$ distance between two adjacent elements, and wave number $k = 2 \cdot \pi/\lambda$ (λ=wavelength).

Values I_n and δ_n are determined by specific design of beam, forming network. Typically, the array factor is expressed by an absolute value (2.1) normalized to its maximum and is plotted in dB scale.

$$\left|AF_{\text{linear}}^{\text{norm}}(\theta,\phi)\right| = \frac{|AF_{\text{linear}}(\theta,\phi)|}{\max\{|AF_{\text{linear}}(\theta,\phi)|\}} \tag{2.2}$$

For the uniform amplitude and equal phase distributions ($I_n = I$ and $\delta_n = \delta$, $n = 1, 2 \ldots N$) normalized array factor is given by

$$\left|AF_{\text{linear}}^{\text{norm}}(\theta,\phi)\right| = \frac{1}{N} \cdot \left| \frac{\sin\left(\frac{N}{2} \cdot k \cdot d \cdot \sin\theta\right)}{\sin\left(\frac{k \cdot d \cdot \sin\theta}{2}\right)} \right| \tag{2.3}$$

It is seen that the linear factor (2.3) does not depend on ϕ value and has a maximum equal to 1 for the angle direction $\theta = 0$. As we can see, function (2.3) also has maximum value for the following angle directions (grating lobe angle directions)

$$\theta_r = \pm \arcsin\left(\frac{\lambda \cdot r}{d}\right); \quad r = 1, 2\ldots \tag{2.4}$$

If the distance between the adjacent elements is equal or less than the wavelength λ, linear antenna array has only one beam peak within the visible observation angle region (−90°–90°). When $d > \lambda$ the unwanted beam peak (grating lobe) occurs in the real angle range of −90°–90°. Therefore, observation angle range dictates the value of the maximum element spacing to avoid the occurrence of the grating lobe. For example, if the observation angle occurs in the range −30° and 30° array element spacing can be chosen as $2 \times \lambda$. As it follows from the formula (2.3), the value of the maximum sidelobe (with respect to the main beam peak) for the array with uniform amplitude distribution is about −13.1 dB, and the angle direction of this lobe can be estimated from the following expression

$$\theta_{\text{max lobe}} = \pm \arcsin\left(\frac{3}{2} \cdot \frac{1}{N} \cdot \frac{\lambda}{d}\right) \tag{2.5}$$

Figure 2.2 shows simulation example results for the normalized linear array factor with different element numbers. All diagrams are obtained for uniform amplitude distribution and normalized to the main lobe value, (normal to the array (direction z) corresponds to the angle 0°) Figs. 2.2a–d are obtained for inter-element spacing equal half wavelength and Fig. 2.2e–h demonstrate the array factor for inter-element spacing equal to one wavelength. As we can see, one wave spacing generates grating lobes with a magnitude equal main lobe value. Values for the maximum side lobes are around −13 dB as it follows from the expression (2.3). The beamwidth of the main lobe between two adjacent nulls is about

$$\Delta\theta_{\text{main}} \approx \frac{2 \cdot \lambda}{d \cdot N} \tag{2.6}$$

The directivity of the broadside linear array factor toward the normal to the array with uniform amplitude and phase distribution is given by [15]

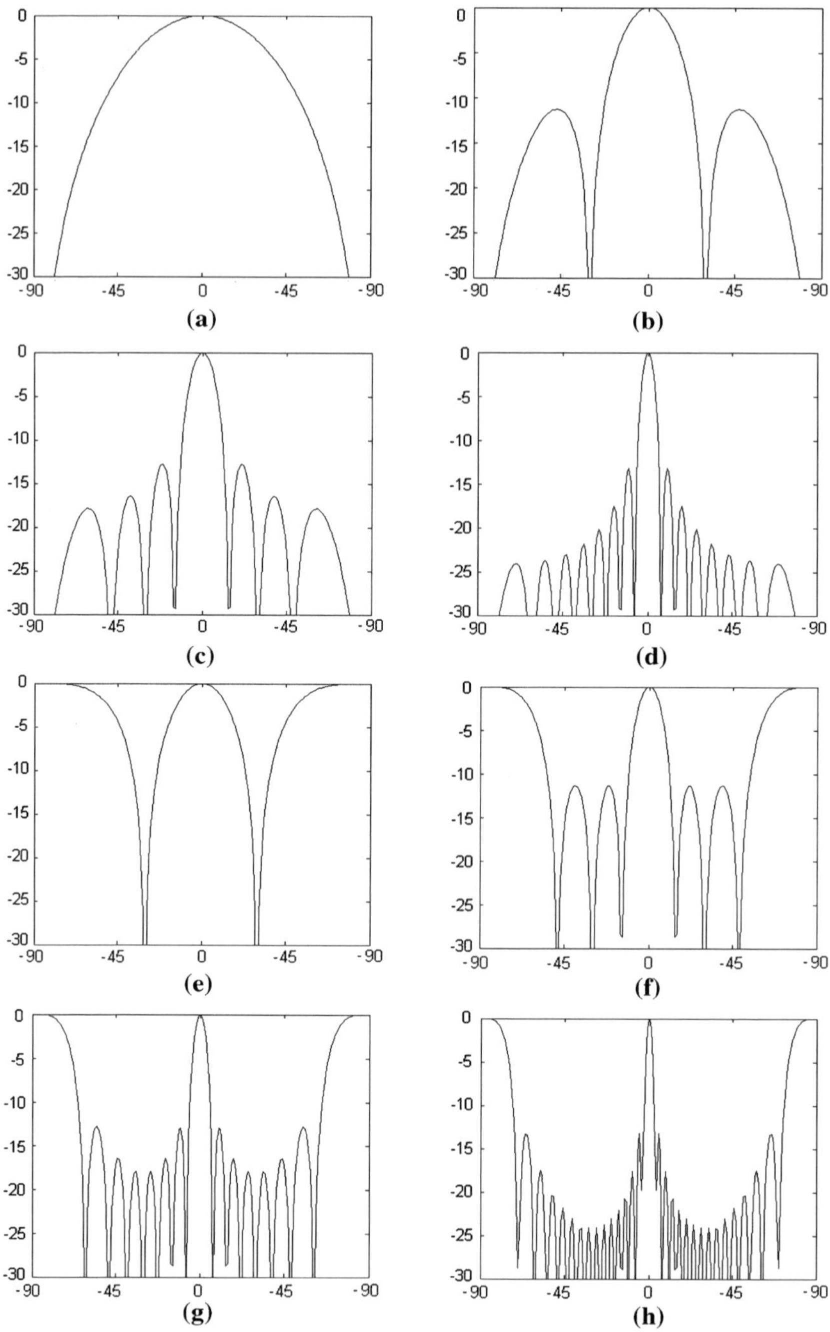

Fig. 2.2 Linear array factor. **a** Two element array, spacing $\lambda/2$; **b** Four element array, spacing $\lambda/2$; **c** Eight element array, spacing $\lambda/2$; **d** Sixteen element array, spacing $\lambda/2$; **e** Two element array, spacing λ; **f** Four element array, spacing λ; **g** Eight element array, spacing λ; **h** Sixteen element array, spacing λ

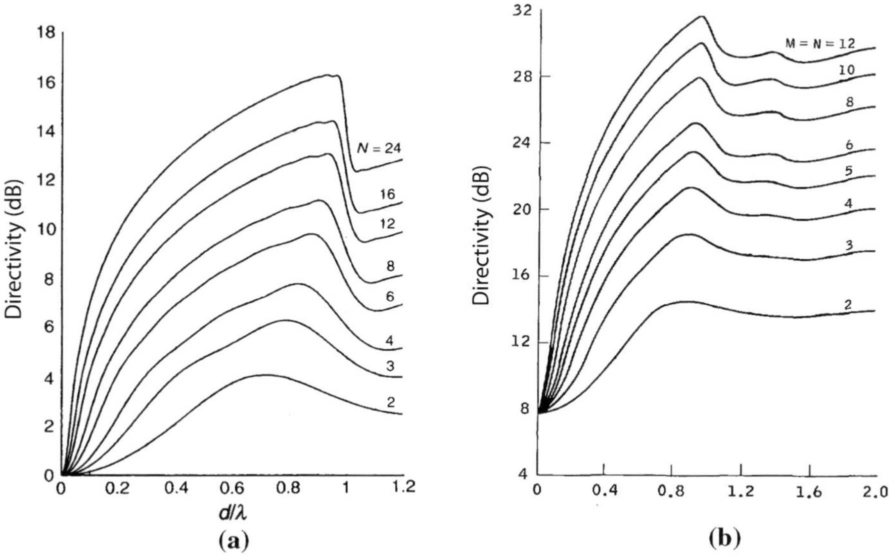

Fig. 2.3 Directivity as a function of the element spacing: **a** Linear array factor; **b** Planar array factor

$$D_{\text{linear}} = \frac{N^2}{N + 2 \cdot \sum_{n=1}^{N-1} \frac{(N-n)\cdot\sin(n\cdot k\cdot d)}{n\cdot k\cdot d}} \tag{2.7}$$

For half wave inter-element spacing, directivity D_{linear} is equal to N. Figure 2.3a shows calculated directivity values as a function of element spacing for different numbers of isotropic elements. It is seen that the directivity graph drops at the appearance of the first grating lobe.

Arrays with feed networks shown in Fig. 1.7c, d produce non-uniform amplitude distribution, which reduces sidelobe level and, therefore, protects the receiving system from interference signals. For example, the maximum sidelobe level can be reduced from -13 dB up to -17 dB (relatively to the main beam peak) if the currents I_n at the edge array elements are about of 0.5 from the maximum value at the central array elements and up to -21 to-40 dB for the edge antenna array element currents equal to 0. The amplitude tapering for the feed network shown in Fig. 1.7c or d can be controlled by the distance between the feed antenna and the array plane or by variation of the feed antenna radiation pattern. Non-uniform amplitude distribution across the array can also be realized with parallel, series, or parallel-series networks [16–18] shown in Fig. 1.5. For example, the reference paper [16] describes six element array with series fed network, which provides amplitude distribution with sidelobe level of about -20 dB. However, the amplitude tapering produces the broadening of the main beam and reduced antenna gain. The broadside directivity of the linear array factor with amplitude distribution I_n and equal phases for all elements is given by [19, 20].

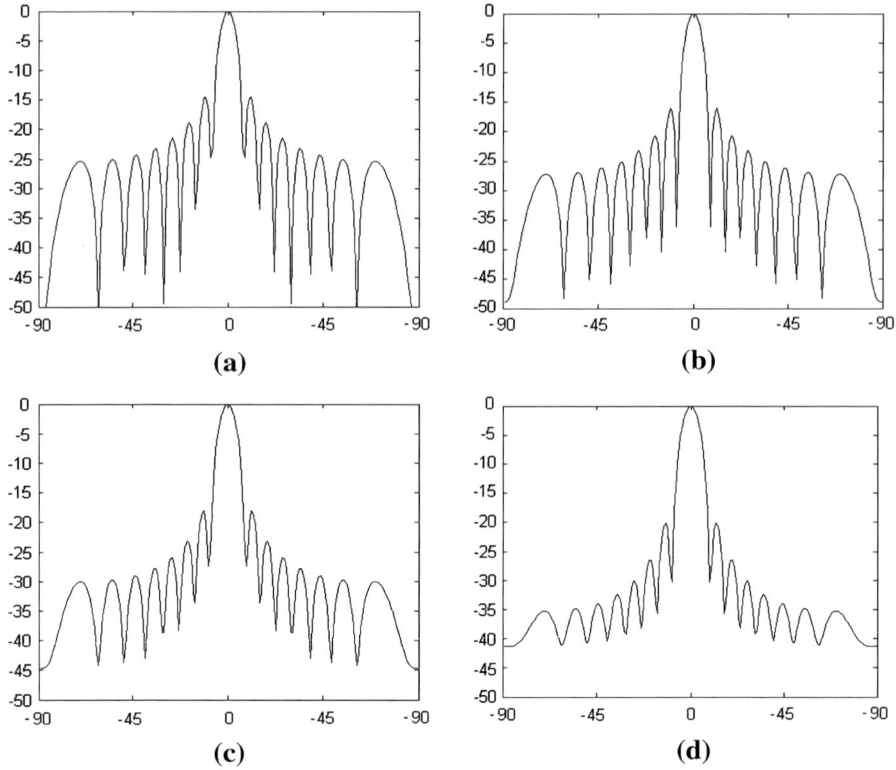

Fig. 2.4 Linear array factor, eight elements, element spacing $\lambda/2$: **a** Uniform amplitude distribution ($\gamma = 0$); **b** $\gamma = 0.4$; **c** $\gamma = 0.6$; **d** $\gamma = 0.8$

Table 2.1 Aperture (taper) efficiency

N	16	8	4
$\eta_{\text{taper}}(\gamma = 0.2)$	0.995	0.99	0.99
$\eta_{\text{taper}}(\gamma = 0.4)$	0.976	0.968	0.95
$\eta_{\text{taper}}(\gamma = 0.6)$	0.936	0.916	0.862
$\eta_{\text{taper}}(\gamma = 0.8)$	0.87	0.828	0.709

$$D_{\text{linear}}^{\text{taper}} = \frac{\left(\sum\limits_{n=1}^{N} I_n\right)^2}{\sum\limits_{n=1}^{N}\sum\limits_{m=1}^{N} I_n \cdot I_m \cdot \frac{\sin((n-m)\cdot k\cdot d)}{(n-m)\cdot k\cdot d}} \tag{2.8}$$

The ratio $\eta_{\text{taper}} = \frac{D_{\text{linear}}^{\text{taper}}}{D_{\text{linear}}}$, named as excitation (taper, or aperture) efficiency, determines the directivity loss due to the tapering amplitude from the center array to the edges. When the distance between the array elements is equal to half of the wavelength, η_{taper} is given by

$$\eta_{\text{taper}} = \frac{\left(\sum_{n=1}^{N} I_n\right)^2}{N \cdot \sum I_n^2} \tag{2.9}$$

For a uniform amplitude distribution, excitation efficiency is equal to 1. If the first sidelobe level of about −18 dB η_{taper} is around 0.95 (−0.22 dB) [20], for first sidelobe value of −23 dB $\eta_{\text{taper}} = 0.81$ (−0.92 dB), and for first sidelobe level equal to −31.5 dB $\eta_{\text{taper}} = 0.67$ (−1.7 dB). Figure 2.4 shows simulation result examples of the linear array factor with the number of elements N = 8, inter-element spacing equal to half of the wavelength, and different non-uniform amplitude distribution functions (normal to the array (direction z) corresponds to the angle 0°).

$$I_n = 1 - \gamma \cdot \left(\frac{n - N/2 - 0.5}{N/2 - 0.5}\right)^2 \tag{2.10}$$

Amplitude vales (2.10) are symmetrically equal, i.e., $I_n = I_{N-n+1}$ ($n = 1, 2\ldots N$) for even number N. Value $\gamma = 0$ corresponds to the uniform distribution and $\gamma = 1$ determines illumination of the array aperture with magnitude I_n which decreases almost to zero from the center to the edge of the array. Table 2.1 presents the taper efficiency calculated with formulas (2.9) and (2.10) for different element numbers N.

The aperture efficiency value <1 shortens communication range. However, reduced sidelobes increase the immunity to the interference sources. Therefore, antenna design engineer has to choose reasonable compromise between the sidelobes level and the gain loss to satisfy the communication system requirements.

2.3 Planar Array Factor

The array factor for the planar equally spaced array with N elements in each column and M elements in each row (Fig. 2.1b) is as in following [21]

$$AF_{\text{planar}}(\theta, \phi) = \sum_{n=1}^{n=N} \sum_{m=1}^{m=M} I_{nm} \cdot e^{j \cdot (\delta_{nm} + k \cdot d \cdot n \cdot \sin\theta \cdot \cos\phi + k \cdot d \cdot m \cdot \sin\theta \cdot \sin\phi)} \tag{2.11}$$

For uniform amplitude distribution ($I_{nm} = 1$) and equal phase distribution ($\delta_{nm} = 0$), the normalized planar array factor is defined as

$$\left|AF^{\text{norm}}_{\text{planar}}(\theta,\phi)\right| = \left|\frac{\sin\left(\frac{N}{2}\cdot k\cdot d\cdot\sin\theta\cdot\cos\phi\right)}{N\cdot\sin\left(\frac{k\cdot d\cdot\sin\theta\cdot\cos\phi}{2}\right)}\right|\cdot\left|\frac{\sin\left(\frac{M}{2}\cdot k\cdot d\cdot\sin\theta\cdot\sin\phi\right)}{M\cdot\sin\left(\frac{k\cdot d\cdot\sin\theta\cdot\sin\phi}{2}\right)}\right|$$
$$= S_N(\alpha)\cdot S_M(\beta) \tag{2.12}$$

where the coordinates α and β are determined as $\sin\alpha = \sin\theta\cdot\cos\phi$, $\sin\beta = \sin\theta\cdot\sin\phi$ and

$$S_N(\alpha) = \left|\frac{\sin\left(\frac{N}{2}\cdot k\cdot d\cdot\sin\alpha\right)}{N\cdot\sin\left(\frac{k\cdot d\cdot\alpha}{2}\right)}\right|,\ S_M(\beta) = \left|\frac{\sin\left(\frac{N}{2}\cdot k\cdot d\cdot\sin\beta\right)}{N\cdot\sin\left(\frac{k\cdot d\cdot\beta}{2}\right)}\right|; \tag{2.13}$$

The main lobe (principal maximum) and grating lobes of the terms $S_N(\alpha)$ and $S_M(\beta)$ are located at the angles (similar to the linear array) such that

$$\frac{d}{\lambda}\cdot\sin\alpha = \pm r_1;\ \frac{d}{\lambda}\cdot\sin\beta = \pm\, r_2;\quad r_1 = 0,1,2\ldots;\quad r_2 = 0,1,2\ldots \tag{2.14}$$

Main lobe corresponds to $r_1 = 0\ (\alpha = 0°)$ and $r_2 = 0\ (\beta = 0°)$. To avoid grating lobes, the row and column spacing between the elements of the planar array has to be less than the wavelength. The beamwidth value of the main lobe between two adjacent nulls along the axis x can be determined from the expressions (2.5) and (2.6) by replacing θ with α and element spacing d with d_x. Similar logic can be used for estimating of the beamwidth value along the axis y by replacing θ with β and d with d_y. Formulas (2.13) present the pattern of planar array as the antenna pattern of two multiplicatively combined orthogonal linear arrays [22, 23]. These expressions show that the angular resolution of the planar array with N^2 elements in the main X or Y planes can be equivalent to the array consisting of two orthogonal linear arrays with total number of elements 2 N. More than 50 years ago, Mills [24] employed such technique (Mills Cross Technique) to achieve high angular resolution with a relatively small numbers of antennas.

Directivity for the planar array factor is expressed by a more complicated formula than Eq. (2.7) or (2.8). However, for the array with a large number of isotropic elements, directivity value can be estimated by the simple rough approximation [21]

$$D_{\text{planar}} \approx \pi\cdot D_{\text{linearx}}\cdot D_{\text{lineary}} \tag{2.15}$$

where D_{linearx} and D_{lineary} are the directivities of linear array factors along the x and y axes, respectively. For a planar array with half wave spacing along the x and y coordinates, directivity value is proportional to the total array element number $D \simeq \pi\cdot N\cdot M$. More accurate value [25, 26] directivity value for different element numbers, as a function of the element spacing are presented in Fig. 2.3b. Plots, like curves obtained for linear array, have dips caused by the emergence of

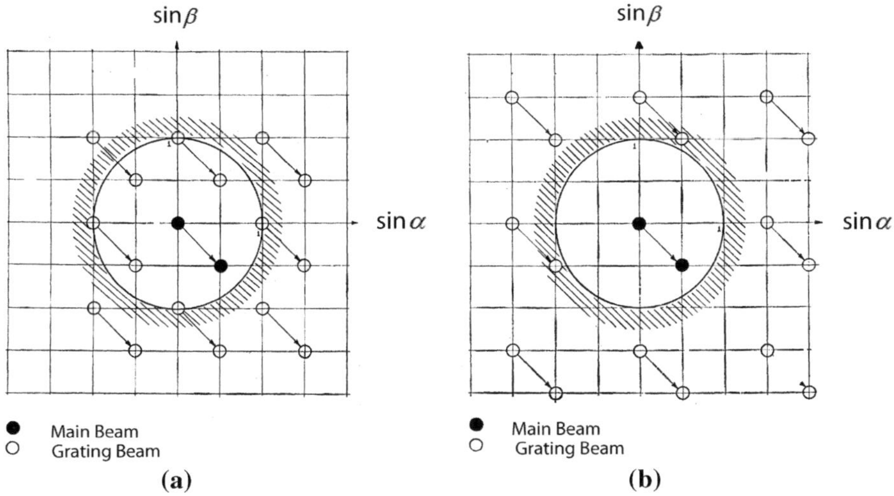

Fig. 2.5 Diagram showing grating beams: **a** Element spacing equal λ; **b** Element spacing equal 2/3 × λ

grating lobes into the visible region when the element spacing is closed to the wavelength.

2.4 Array Factor for Circular Configuration

Referring to Fig. 2.1d, let us assume that N equally spaced isotropic elements are placed on x–y plane along a circular ring of radius a. The antenna factor can be expressed as [21]

$$AF_{\text{circlar}} = \sum_{n=1}^{N} I_n \cdot e^{j\cdot(k\cdot a\cdot\sin\theta\cdot\cos(\phi-\phi_n)+\delta_n)} \tag{2.16}$$

where I_n = amplitude excitation of the nth element, $\phi_n = \frac{2\cdot\pi\cdot n}{N}$ = nth angular position of element, δ_n = phase excitation of the nth array element. The maximum of the AF occurs when all the phase terms in (2.16) are equal

$$k\cdot a\cdot\sin\theta\cdot\cos(\phi-\phi_n)+\delta_n = 2\cdot\pi\cdot r \quad r = 0, \pm1, \pm2\ldots \tag{2.17}$$

If $\delta_n = 0$ for all n numbers, the main maximum ($r = 0$) occurs for the angle direction θ=0°

For uniform amplitude distribution ($I_n = 1$ for $n = 1, 2 \ldots N$) expression (2.16) can be simplified as $AF_{\text{circlar}} \approx N\cdot J_0(k\cdot\rho_0)$; where $J_0(x)$ = zero order Bessel function; $\rho_0 = a\cdot\sin\theta$. Typically, circular arrays are used in the smart antenna systems that provide adaptive beamforming to enable the formation (simultaneously or successively) of beam toward the desired signal angle direction and nulls toward the angles of interfering signals. Uniformly circular antenna array can be electronically rotated in the plane of array without significant change of the beam shape, providing 360° azimuth coverage area [27].

2.5 Array Factor for Electronically Controlled Phased Arrays

2.5.1 Design with Analog Phase Shifters

Schematically, linear passive phase array with electronic beam control is presented in Fig. 1.7a. The linear system consists of N equally spaced identical isotropic elements. Each phase shifter has a special electrical control circuit, that can change the phase of the received signal. Let us assume that the space between the elements of the equally spaced linear antenna array is equal to d, and we want to receive a signal coming from the angle direction θ_0. Assume that electronically controlled phased shifters provide a progressive phase shift between the adjacent antenna elements $\Delta\delta = \Delta\delta(\theta_0)$, then

$$\Delta\delta(\theta_0) = -k \cdot d \cdot \sin\theta_0;\ \delta_n = -k \cdot d \cdot n \cdot \sin\theta_0; \tag{2.18}$$

The linear antenna array factor (2.1) with uniform amplitude distribution can be expressed as

$$|AF_{\text{linear}}(\theta,\ \theta_0)| = \left| \frac{\sin\left(\frac{N}{2} \cdot k \cdot d \cdot (\sin\theta - \sin\theta_0)\right)}{\sin\left(\frac{k \cdot d \cdot (\sin\theta - \sin\theta_0)}{2}\right)} \right| \tag{2.19}$$

Equation (2.19) has a maximum value equal to N for the angle directions $\theta = \theta_r$

$$\theta_r = \arcsin\left(\pm \frac{\lambda \cdot r}{d} + \sin\theta_0 \right);\ r = 0, 1, 2\ldots \tag{2.20}$$

$r = 0$ corresponds to the main beam angle position ($\theta = \theta_0$) and $r \neq 0$ determines the angle positions of the grating unwanted lobes.

As we can see from the expression (2.20), the closest unwanted grating lobe will be not appeared in visible space when

$$\frac{d}{\lambda} < \frac{1}{1 + \sin|\theta_0|} \tag{2.21}$$

Let us assume that that scanning angle range is equal $|\theta_0| < 60°$. Expression (2.21) indicates that inter-element space d in this case is $<0.53\lambda$. Therefore, to avoid grating lobs in scanning angle sector −60 to +60 degrees we have to choose an inter-element spacing of about 0.53λ, not more.

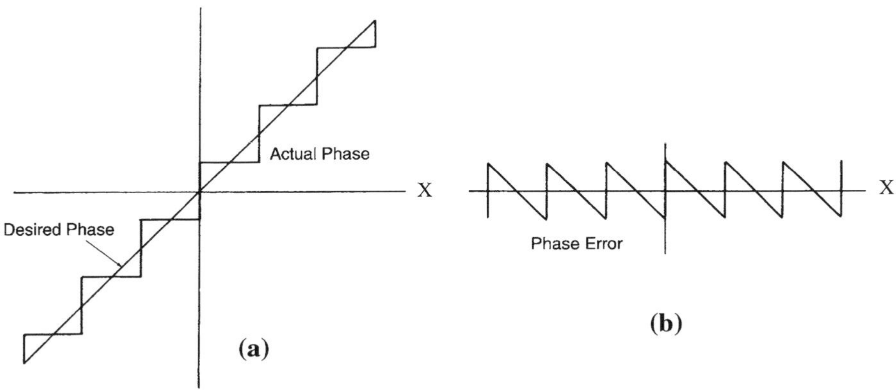

Fig. 2.6 **a** Phase shift for electronic scanning; **b** Phase error function

The absolute value of array factor for equally spaced planar array is given by

$$|AF_{\text{planar}}| = \left|\frac{\sin\left(\frac{N}{2}\cdot k\cdot d\cdot(\sin\alpha-\sin\alpha_0)\right)}{\sin\left(\frac{k\cdot d\cdot(\sin\alpha-\sin\alpha_0)}{2}\right)}\right|\cdot\left|\frac{\sin\left(\frac{M}{2}\cdot k\cdot d\cdot(\sin\beta-\sin\beta_0)\right)}{\sin\left(\frac{k\cdot d(\sin\beta-\sin\beta_0)}{2}\right)}\right| \tag{2.22}$$

Where $\sin\alpha = \sin\theta\cdot\cos\phi$; $\sin\alpha_0 = \sin\theta_0\cdot\cos\phi_0$; $\sin\beta = \sin\theta\cdot\sin\phi$; $\sin\beta_0 = \sin\theta_0\cdot\sin\phi_0$

To avoid grating lobes for scanning angles $\alpha_0 = \pm 60°$ or $\beta_0 = \pm 60°$ from broadside direction it is necessary to choose inter-space distances along x and y coordinates $<0.53\lambda$.

Figure 2.5 presents the diagram which shows the number of grating beams in a visible range, depending on the scanning angle values α and β.

Figure 2.5a corresponds to the spacing between elements equal wavelength and Fig. 2.5b is obtained for spacing equal to two-thirds of wavelength. Equation (2.22) is an estimation of the radiation pattern with analog phase shifters that vary phases continuously from 0° to 360°. Such electronically controllable phase shifters are very expensive and typically are not used in practice.

2.5.2 System with Digital Phase Shifters

Most phase shifters are digitally controlled, so they realize phase shifts with a discrete Δ equal to $\Delta = \frac{2\cdot\pi}{2^q}$, where q is the number of bits, and 2^q is the number of digital phase shifter phase states. For example, a one-bit (q=1) digital phase shifter produces only two phases: 0° and 180°, a two-bit digital shifter can realize four phases 0, $\pi/2$ π, 3 $\pi/2$, and a three-bit digital phase shifter can realize phases:

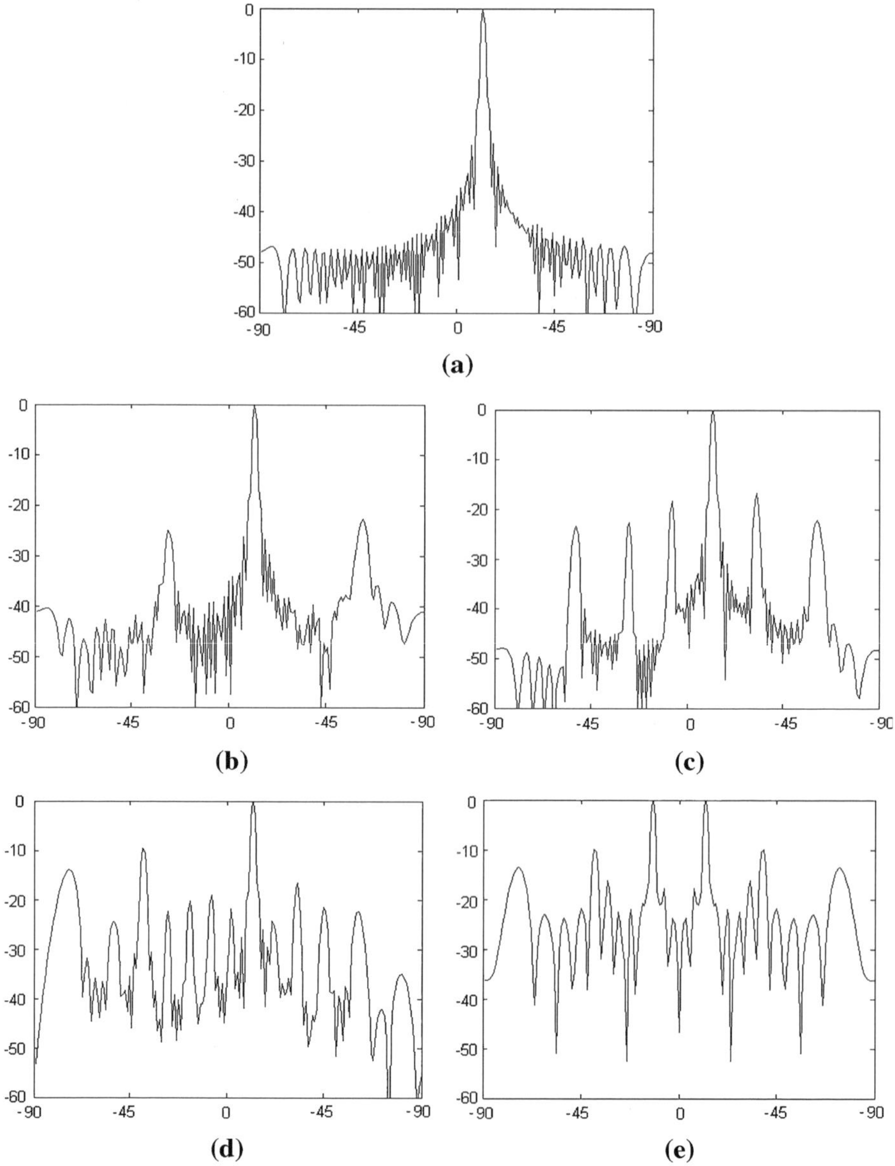

Fig. 2.7 Computer simulation of 64 element linear array with spacing $\lambda/2$, scan angle 12°: **a** Analog phase shifters; **b** $\Delta = 22.5°$; **c** $\Delta = 45°$; **d** $\Delta = 90°$; **e** $\Delta = 180°$

0, $\pi/4$, $\pi/2$, 3 $\pi/4$, π, 5 $\pi/4$, $\pi/2$, 7 $\pi/4$. Figure 2.6a shows the ideal (desired) linear phase curve for electronic scanning as a function of the x coordinate (linear array is placed along this axis) and approximation (actual phase) caused by the stair-step phase. Phase error is demonstrated in Fig. 2.6b.

It is seen that the error between the ideal curve and its approximation is a periodic function of the X coordinate. Periodic errors cause main lobe attenuation, produce a set of lobes called as "quantization lobes", and cause error in the main beam pointing position. It is known [28, 29] that the array factor for a linear antenna array with digital phase shifters can be presented as

$$AF_{\text{linear}}^{\text{digit}} = \sum_{m=-\infty}^{m=\infty} C_m \cdot \sum_{n=1}^{N} I_n \cdot e^{j\cdot\left(k\cdot d\cdot\sin\theta - k\cdot d\cdot\left(\sin\theta_0 + m\cdot\frac{2\cdot\pi}{\Delta}\right)\right)} \tag{2.23}$$

where parasitic lobe coefficients C_m are given by

$$C_m = \frac{(-1)^m \cdot \sin\left(\frac{\Delta}{2}\right)}{\left(\frac{\Delta}{2} - \pi\cdot m\right)} \tag{2.24}$$

Formula (2.23) shows that the array factor of the array with digital phase shifters is a sum of the linear array factors with amplitude weightings (2.24) and beam angle directions

$$\theta_{rm} = \arcsin\left(\frac{\lambda\cdot r}{d} + \left(1 + \frac{2\cdot\pi}{\Delta}\cdot m\right)\cdot\sin\theta_0\right) \tag{2.25}$$

where r, $m = 0, \pm 1, \pm 2 \ldots$ (any positive or negative integer numbers).

An array factor with $m = r = 0$ corresponds to the main beam while beams with number $m \neq 0$ determine unwanted (parasitic or quantization lobes) lobes. It is seen that the angle directions (2.25) depend on the scanning angle θ_0, and the parasitic lobe amplitudes decrease with increasing number m. Simple estimation of the main beam loss effect due to digital phase shifters is

$$\Delta AF_{\text{linear}} \approx C_0 = \frac{\sin\left(\frac{\Delta}{2}\right)}{\frac{\Delta}{2}} \tag{2.26}$$

Quantization lobe (QL) values depend only on the scanning angle position and do not depend on the array amplitude distribution. The maximum value of quantization lobe QL ($m = 1$) is equal [with respect to the main lobe (2.26)] to

$$QL \approx \left|\frac{C_1}{C_0}\right| = \frac{1}{\frac{2\cdot\pi}{\Delta} - 1} \tag{2.27}$$

Figure 2.7 presents computer calculations of the linear array factor for the equally spaced (d/λ=0.5) elements (N = 64). Scan angle θ_0 from the normal to the array is equal to 12°, where normal to the array (direction z) corresponds to the angle 0°. Amplitude distribution provides sidelobe levels <−20 dB. Figure 2.7a presents the results for analog phase shifters, Fig. 2.7b–e show the array factor for the same antenna with digital phase shifters. As we can see, a linear array factor with one bit phase shifters leads to the being split beam into two equally main

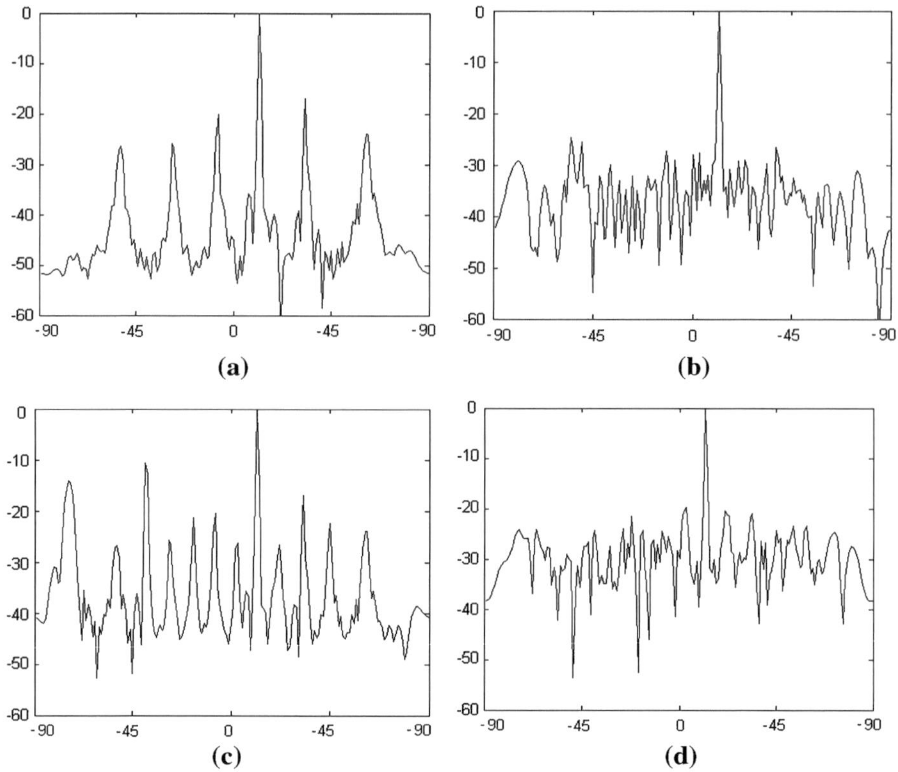

Fig. 2.8 Simulation results for 132 element linear array factor with randomizing periodic phase error: **a** Three-bit phase shifters without randomizing; **b** Three-bit phase shifters with randomizing; **c** Two-bit phase shifters without randomizing; **d** Two-bit phase shifters with randomizing

Table 2.2 Possible phase states for one bit phase shifters

Element Number → Scan Angles ↓	1	2	3	4	5	6	7	8
1: 0°–9°	0	0	0	0	0	0	0	0
2: 9°–12°	π	0	0	0	0	0	0	π
3: 12°–20°	π	π	0	0	0	0	π	π
4: 20°–26°	π	π	π	0	0	π	π	π
5: 26°–37°	0	π	π	0	0	π	π	0
6: 37°–46°	0	0	π	0	0	π	0	0
7: 46°–60°	π	0	π	0	0	π	0	π

lobes. The calculated values of the levels and angle positions of parasitic lobes approximately correspond to the estimations (2.24)–(2.27).

A few different methods [28–32] were offered to reduce these parasitic lobes. One of the simplest and effective methods can be realized using the feed network

Table 2.3 Average linear array factor parameters

q (bit number) →	2	3	4	5
$\delta\theta_0$	1.76	0.98	0.46	0.23
SL_{max}(dB)	−6.97	−10.19	−11.62	−12.18
D_{loss} (dB)	0.82	0.2	0.047	0.012

shown in Fig. 1.7c, d. An antenna array with digital phase shifters is illuminated by a spherical wave with the phase difference between edge and central elements of about 360°. According to the known array geometry, it can calculate phase shifts required for converting spherical wave into a plane wave-front and to calculate linear phase shift along the array aperture for scanning beam toward the predetermined angle direction. In this case, the periodic error between the illuminated phase and the realized with digital phase shifters phase [32] is destroyed, and parasitic lobes are significantly reduced. Similar effect can be obtained if the illuminated wave is presented by quasi-random phase distribution along the array (for example, uniform random phase distribution in the range −180°–180°). Such phase distribution can be produced by special feeding network design. Random phases required for the compensation of feeding network phases are stored in the computer memory and do not depend on the scan angle. Figure 2.8 demonstrates simulation results for the linear array factor of the antenna array consisting of 132 elements under condition that array phase error shown in Fig. 2.6a is destroyed using random phases, and a scan angle of 12°. Figure 2.8a, b demonstrate linear array factor with and without randomizing phase error for three-bit phase shifters, Fig. 2.8c, d show similar curves for two-bit phase shifters. As we can see, randomizing the periodic phase error significantly reduces the parasitic lobes, while increases the average power sidelobe level P_{average}

$$P_{\text{average}} \approx \frac{\Delta^2}{12 \cdot N} \tag{2.28}$$

This value is a rough estimation and usable for multi-element antenna array.

The error in the main beam pointing of the multi-element array can be estimated as [15]

$$\delta\theta \approx \frac{\Delta}{k \cdot d \cdot N \cdot \sqrt{N} \cdot \cos(\theta_0)} \tag{2.29}$$

Where θ_0 = scanning angle.

For small element arrays, the expression (2.28) has to be revised by computer simulations.

As an example, Table 2.2 [33] shows detailed phased distributions for the 8 element array with an inter-element space $\lambda/2$ and one bit digital phase shifters.

With one bit phase shifter, the phase is limited to two states: 0° and 180°. The maximum deviation from the phase that is required for the antenna element to steer

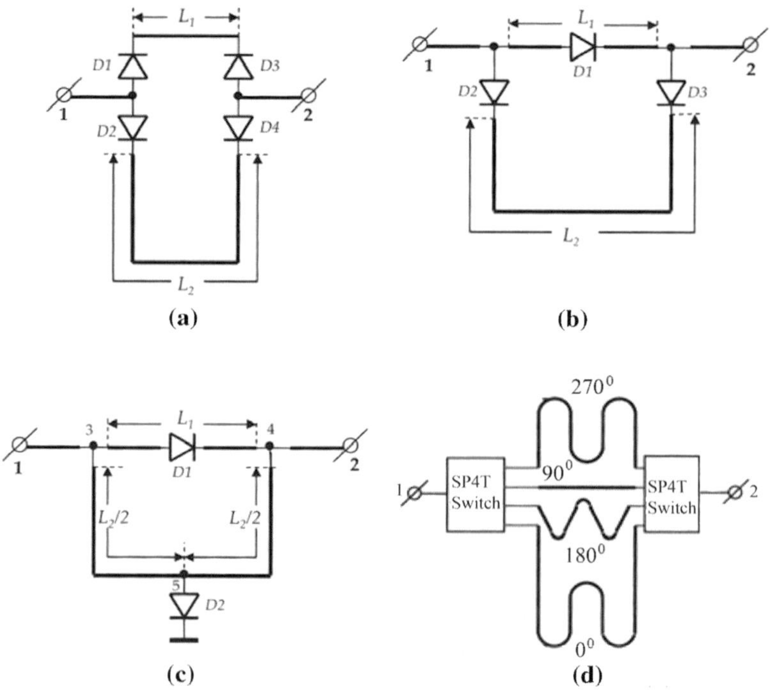

Fig. 2.9 Phase shifter circuit topology: **a–c** Transmission line design; **d** SP4T switch configuration

the beam is 90°. Only seven phase distributions are available to cover 60° scan range.

As we can see, all array elements are in phase for the scan angles from 0°to 9°, and there is no beam steering within this angle range. Note that the maximum power value cannot be reached for $\theta_0 > 9°$ because of the presence of two main lobes symmetrically located with respect to the broadside direction.

A two-bit phase shifter ($q = 2$) provides four phase states within 0–360° phase range: 0°, 90°, 180°, and 270°. The maximum deviation required for each element to steer the beam is then 45°. The array factor value does not change for scan angles from 21° to 30° and from 30.5° to 40.5°.

A three-bit phase shifter ($q = 3$) gives eight possible phase states, starting from 0° to 45°. The maximum deviation of the phase of each element is then 22.5°. The beam pointing remains constant for the largest angular ranges: 43°–49° and 49°–55.5°.

A four-bit phase shifter corresponds to sixteen phase steps with a discrete equal to 22.5°. The maximum deviation of the phase is now 11.25°. Table 2.3 summarizes the average array factor parameters for the eight element array with two- to five-bit phase shifters. The average was realized by considering a scan angle varying from 0° to 60° with a 0.5° step. The scan angle deviation $\delta\theta_0$ is expressed in degrees, SL_{max} is the maximum of sidelobe level, and D_{loss} is an average directivity loss.

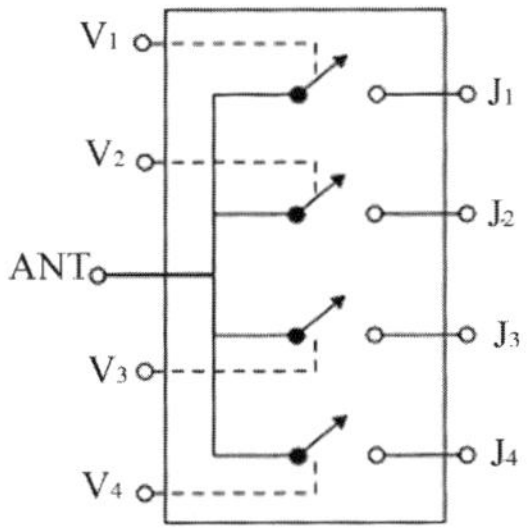

Truth Table

V_1	V_2	V_3	V_4	Ant-J_1	Ant-J_2	Ant-J_3	Ant-J_4
V_{High}	V_{Low}	V_{Low}	V_{Low}	Ins. loss	Isolation	Isolation	Isolation
V_{Low}	V_{High}	V_{Low}	V_{Low}	Isolation	Ins. loss	Isolation	Isolation
V_{Low}	V_{Low}	V_{High}	V_{Low}	Isolation	Isolation	Ins. loss	Isolation
V_{Low}	V_{Low}	V_{Low}	V_{High}	Isolation	Isolation	Isolation	Ins. loss

(a)

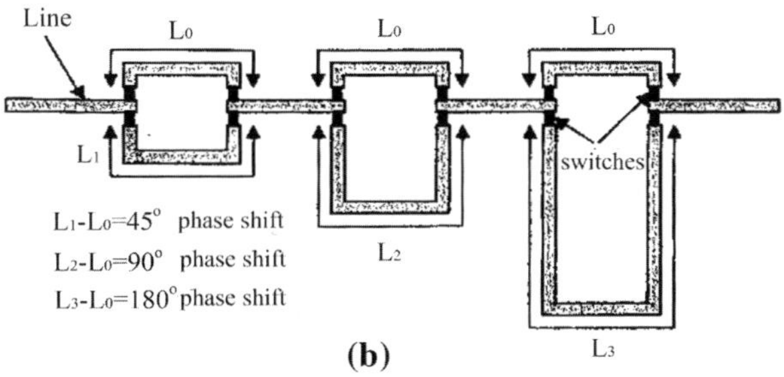

(b)

Fig. 2.10 **a** Truth table for SP4T switch; **b** Three—Bit switched line phase shifter

2.5.3 Pin Diode Digital Phase Shifter Topology

The phase of an electromagnetic wave is changed when propagating through the channel, which includes phase shifter device. The major parameters of a phase shifter are frequency range, bandwidth, total phase variance, insertion loss, and switching speed. The simplest phase shifter circuits based on transmission line design and pin diodes elements are presented in Fig. 2.9.

By switching the signal between two lines L1 and L2, it is possible to realize specific phase shift [34]:

$$\Delta\phi = \left(\frac{2\cdot\pi\cdot(L2 - L1)}{\lambda}\right) \tag{2.30}$$

Switched line phase shifters generally are digital elements with a discrete $\Delta = \frac{2\cdot\pi}{2^q}$. For example, to get a 180° phase shift, the required physical length difference $\Delta L = L2 - L1$ should be $\lambda/2$, to provide a phase shift equal to 90°, it is necessary to have a physical length difference of $\lambda/4$, and for a 45° phase $\Delta L = \lambda/8$. Typically, digital phase shifters use PIN diodes as phase control elements due to their high speed switching time (a few microseconds) and relatively simple bias circuits. PIN diodes insertion losses are about 1–3 dB, depending of the frequency range. Commercially available pin diode switch assemblies are classified by the different numbers of outputs: SPST (one input, one output), SPDT (one input, two outputs), SP3T (one input, three outputs) etc. Pin diode Truth Table shows controllable voltage values (High or Low) when connecting the input line with the output line to provide the desired phase shift. The block diagram example of a two bit phase shifter based on use of SP4T switch is demonstrated in Fig. 2.9d, and Truth Table for such switch is presented in Fig. 2.10a.

The control voltage V_{High} is equal 5 volts and V_{Low} is zero volts. The isolation corresponds to the losses of 30 dB and the insertion losses are <1.5 dB. Figure 2.10b shows a two-bit switched line phase shifter using SPDT integrated circuits. Many commercially pin diode switches are available on the market, for example, from Skyworks Solutions, RF Micro Devices, Avago Technologies, etc.

2.6 Antenna Array Performances

2.6.1 Array Element Radiation Pattern and Mutual Coupling

So far, we supposed that the array elements have omnidirectional radiation patterns. The elements in a real array are not isotropic or isolated sources. The array element radiation pattern [35] is determined as a pattern taken with a feed at a single element in the array, and all other elements are terminated by the matched loads. The pattern $F_{m\,\text{element}}(\theta, \phi)$, called an "active element pattern", is different from the isolated element pattern $F^{\text{isolated}}_{\text{element}}(\theta, \phi)$, which reveals the radiation pattern of the antenna element in free space without coupling from "neighbors". An active element pattern depends on the position of the element in the array: patterns of edge elements differ from the center located antenna pattern.

The finite array coupling effects may produce increased sidelobes, main beam squint, shifted nulls, and array blindness in some scan angles. Therefore, it is very important to study the antenna array parameters, including mutual coupling effects. An active element radiation pattern surrounded by others elements can be described using coupling scattering coefficients S_{mn} [37]

$$F_{m\,\text{element}}(\theta) = F^{\text{isolated}}_{\text{element}}(\theta) \cdot \left(C_{mm} \cdot e^{j \cdot k \cdot d \cdot (m-1) \cdot \sin\theta} + \sum_{n=1;\, n \neq m}^{N} S_{nm} \cdot e^{j \cdot k \cdot d \cdot (n-1) \cdot \sin\theta} \right) \tag{2.31}$$

Coefficient C_{mm} denotes the coupling from the array aperture to the output transmission line.

The active reflection coefficient seen at the mth, element is given as [38]

$$\Gamma_m(\theta) = e^{j \cdot k \cdot d \cdot (m-1) \cdot \sin\theta} \cdot \sum_{n=1}^{N} S_{mn} \cdot e^{-(n-1) \cdot k \cdot d \cdot \sin\theta} \tag{2.32}$$

Generally, $\Gamma(\theta) \neq \Gamma(-\theta)$, except of an infinite array of symmetric elements or for the central element of a finite array having an odd number of symmetric elements.

The results of the fundamental reference paper [35] shows that the gain of the active element pattern in the large array where end effects can bc ignored is given by simple formula

$$G_{\text{element}}(\theta) = G_{\text{isolated}}(\theta) \cdot \left[1 - |\Gamma(\theta)|^2\right] \tag{2.33}$$

where $\Gamma(\theta) = \Gamma_m(\theta) = \Gamma_m(-\theta)$ ($m = 1, 2 \ldots N$), $G_{\text{isolated}}(\theta) =$ gain of a single isolated element.

The upper limit (maximum) energy value received by the single element is proportional to the array element area. Furthermore, since the effective area of an element is proportional to its projected area $A = d_x \cdot d_y$ (d_x and d_y are element spacings along x and y axis) in the direction of interest, the gain should have a $\cos\theta$ variation with an angle

$$G^{\max}_{\text{element}}(\theta) = \frac{4 \cdot \pi \cdot d_x \cdot d_y}{\lambda^2} \cdot \cos\theta \tag{2.34}$$

The active reflection coefficient $\Gamma_m(\theta_0)$ can be used to compute the active input impedance at the mth array element [38]

$$Z^m_{in}(\theta) = Z_0 \cdot \frac{1 + \Gamma_m(\theta)}{1 - \Gamma_m(\theta)} \tag{2.35}$$

where $Z_0 =$ is the characteristic impedance of the transmission line with matched generator.

If the scanning angle is equal to zero, then

$$Z^m_{in}(0) = Z_0 \cdot \frac{1 + \sum\limits_{n=1}^{N} S_{mn}}{1 - \sum\limits_{1}^{N} S_{mn}} \tag{2.36}$$

The matrix [S] for the scattering parameters and matrix [Z] for the impedance parameters are related as [15]

$$[S] = \frac{[Z] - Z_0}{[Z] + Z_0} \tag{2.37}$$

Impedance matrix [Z] contains all the inter-element mutual impedances Z_{mn}; the mutual impedances, in principle, are calculated between two elements, with all the other elements open circuited. The mutual impedance between any two elements in the array is found by dividing the open-circuit voltage at one element by the current at the other element

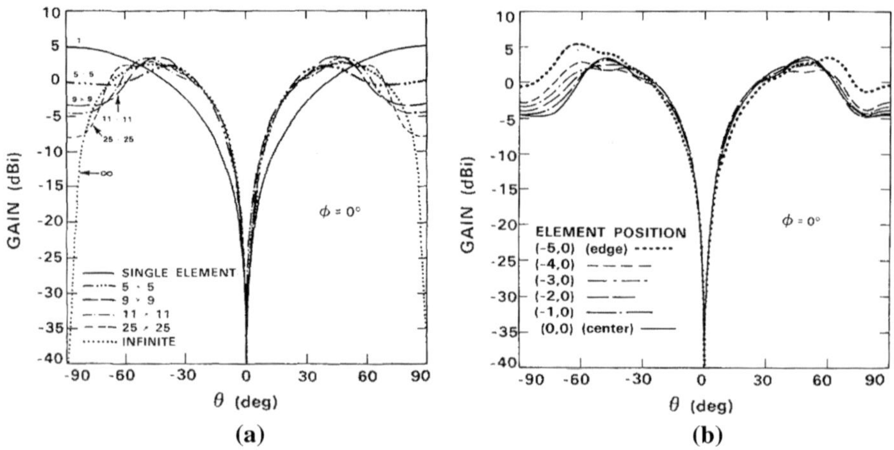

Fig. 2.11 **a** Radiation pattern simulation results for monopole element of planar array:Central element, different number of array elements; **b** Radiation pattern versus element position; [39] ©1985 IEEE

Table 2.4 Reactive impedance of central element versus array size

No. rows #	No. columns	Total number of elements	Reactive portion
3	3	9	+j5.9
5	5	25	+j3.2
7	7	49	+j2.4
9	9	81	+j2.1
11	11	121	+j2.0
25	25	625	+j1.8

$$Z_{mn} = \frac{V_m}{I_n} \text{ for } I_m = 0 \tag{2.38}$$

For an array consisting of two elements, the relationships between mutual scattering components and mutual impedance components are given by

$$S_{12} = \frac{2 \cdot Z_{12}}{(Z_{11}+1)\cdot(Z_{22}+1) - Z_{12}\cdot Z_{21}} \tag{2.39}$$

$$S_{21} = \frac{2 \cdot Z_{21}}{(Z_{11}+1)\cdot(Z_{22}+1) - Z_{12}\cdot Z_{21}} \tag{2.40}$$

$$S_{11} = \frac{(Z_{11}-1)\cdot(Z_{22}+1) - Z_{12}\cdot Z_{21}}{(Z_{11}+1)\cdot(Z_{22}+1) - Z_{12}\cdot Z_{21}} \tag{2.41}$$

Table 2.5 Reactive impedance as a function of element location for 11 × 11 antenna array

Element position	Imaginary impedance portion (ohms)
Center element (0,0)	+j2
Edge element of the center row	+j7
Corner element	+j12
Isolated element (Reference)	+j21

Table 2.6 Amplitude of the coupling coefficient (dB scale) between central array element and element number n ($n = 1, 2, 5$)

Element #	1	2	3	4	5
$\|S_{0n}\|^2$ (dB)	−10	−22	−32	−38	−43

$$S_{22} = \frac{(Z_{22} - 1) \cdot (Z_{11} + 1) - Z_{12} \cdot Z_{21}}{(Z_{11} + 1) \cdot (Z_{22} + 1) - Z_{12} \cdot Z_{21}} \tag{2.42}$$

If the antenna elements are reciprocal, then

$$S_{11} = S_{22} \text{ and } S_{12} = S_{21}\, Z_{12} = Z_{21} \text{ and } Z_{11} = Z_{22} \tag{2.43}$$

Below, we present some simulation and experiment results for the "active radiation pattern" for the element that is surrounded by the other passively terminated elements.

We collect and summarize the results for two kinds of arrays: monopole systems which are good candidates for car applications because they can be easily mounted on the car roof, and microstrip patch arrays for communication between roadside base stations and the vehicle.

a) Monopole Antenna Array

Figure 2.11a [39, 40] demonstrates simulation results for the central element of the planar array with different number of the radiators.

An array consisting of the quarter wavelength monopole elements is placed on the infinite ground plane (xy plane shown in Fig. 2.1b), and the element spacing is equal to half of the wavelength. The monopoles are assumed electrically thin, with a wire radius of 0.001 λ. The radiation pattern is simulated for the single isolated monopole and for the actively exited monopole surrounded by 50 ohm terminated monopoles. Results are presented for the E_θ radiation pattern component in elevation plane, ground plane has infinite size. For the planar array with 5 columns and 5 rows the principal peak occurs at $\theta = 38°$, for the arrays 9 × 9, 11 × 11, and 25 × 25 peak element gain is close to $\theta = 50°$. It is seen, that for 11 × 11 array the element radiation pattern has small differences when compared to the results obtained for the 25 × 25 array. It means that the coupling coefficients S_{mn} (expression (2.31)) become negligible for the distance between the elements of

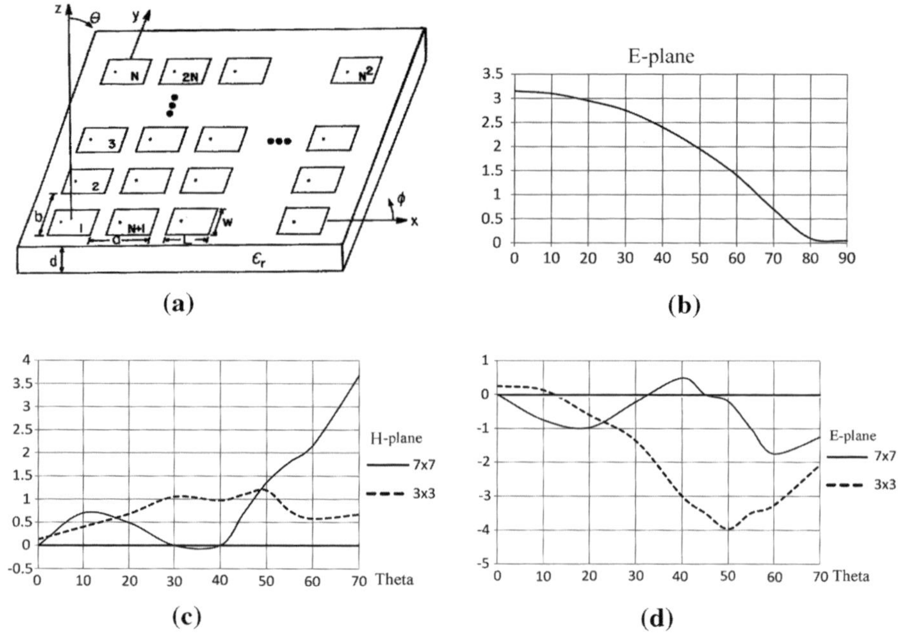

Fig. 2.12 **a** Rectangular patch array; **b** Radiation pattern of the central element in E-plane; **c**, **d** Variations of the radiation pattern of the central array element as a function of array element number

Table 2.7 Coupling coefficient for microstrip array as a function of spacing between elements

Spacing in wavelength	0.25	0.5	0.75	1	1.25
$\lvert S_{12}\rvert^2$ (dB)	−21	−23	−24	−25	−25

about seven wavelength. Also, the center element input impedance versus array size is summarized in Table 2.4. Input impedance for isolated $\lambda/4$ monopole is equal to 36 + j21 Ohm. The real part of the input impedance of the active element is about 38 ohm and practically does not depend on the array size. However reactive impedance portion decreases when the number of array elements increases as shown in Table 2.4. So, impedance of the $\lambda/4$ monopole with $\lambda/2$ spacing tends to be purely resistive as the number of surrounding loaded monopoles increases, and the resistance is insensitive to the array size.

Figure 2.11b demonstrates the element pattern gain as a function of the element position along the center row of the 11 by 11 monopole array. Approximately, 2.2 dB of asymmetry occurs between the two (left and right) peaks of the edge element pattern. For monopoles, two or more elements away from the edge element, the pattern peak asymmetry is <0.5 dB. Table 2.5 shows the input impedance of a few elements of the 11×11 element array.

As it is seen, the imaginary part of the input impedance varies by <10 ohm. Calculations show that the real input impedance portion is relatively insensitive to element position in the 11 × 11 element array and is about 39 ohm.

Table 2.6 presents the experimentally measured coupling scattering parameters S for the central row elements in a planar antenna array (11 × 1 elements) operating at 1.3 GHz. Data are shown for the coupling coefficient $|S_{0n}|^2$ between the central active monopole marked as number 0 and nth element in row. The distance between the adjacent elements is about half of the wavelength.

As we can see, the coupling between adjacent elements is about −10 dB, while the coupling value between the central and fifth element is <−40 dB.

b) Rectangular Patch Array

It is known that microstrip antenna arrays [41–43] have found wide application for communication between the roadside base stations and the car antenna due to its several advantages, such as low profile, light weight, and low cost, etc.

Figure 2.12a shows an example of the microstrip finite antenna array model that is used to estimate the radiation pattern for the array element surrounded by a number of terminated to 50 ohm identical radiators. The operating frequency is 1.41 GHz, W = 10.6 cm, L = 6.55 cm, d = 0.16 cm, spacing between elements is equal to $0.5 \cdot \lambda_0$, λ_0=wavelength in free space, dielectric constant is equal to 2.55. Presented data give an idea of how the mutual coupling coefficient depends on the distance between the adjacent microstrip elements.

Figure 2.12b presents central "active" element gain in the E-plane for the infinite array. The radiation pattern in the H-plane is slightly different from the curve demonstrated in Fig. 2.12b. Figure 2.12c, d show active element gain variations (dB scale) in E and H planes for central element in the passively loaded 50 ohm rectangular patch array schematically shown in Fig. 2.12a. Data are calculated based on the reference paper [41]. Solid horizontal line corresponds to the infinite array. Table 2.7 demonstrates the experimental data of the mutual coupling coefficient $|S_{12}|^2$(dB scale) between two adjacent elements as a function of the distance S for the array shown in Fig. 2.12a.

2.6.2 Array Radiation Pattern

If the equally spaced array consists of a large number of antenna radiators, most of the elements "see" the uniform neighboring environment, and $F_{n\,\text{element}}(\theta, \phi)$ can be approximated by an equal for all elements array radiation pattern $F_{n\,\text{element}}(\theta, \phi) = F_{\text{element}}(\theta, \phi)$. In this case, fully exited array radiation pattern can be expressed as the product of the active element radiation pattern and the array factor

$$F_{\text{array}}(\theta, \phi) = F_{\text{element}}(\theta, \phi) \cdot AF_{\text{array}}(\theta, \phi) \tag{2.44}$$

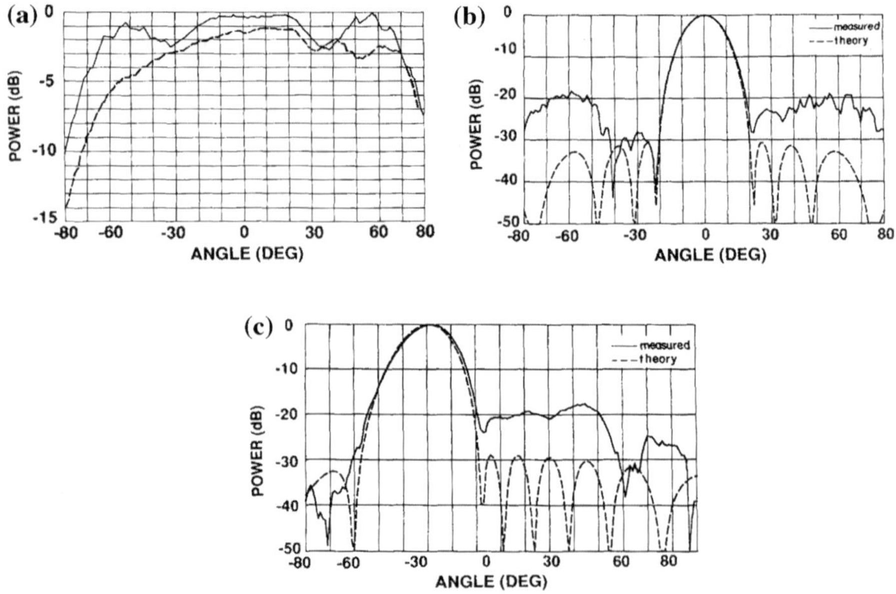

Fig. 2.13 **a** Radiation pattern of central (*solid line*) and edge (*dashed line*) elements of the linear waveguide X-band array; **b** Synthesized 30 dB Chebyshev array radiation patterns without mutual coupling (theory-*dashed line*) and measured with mutual coupling (*solid line*); **c** Radiation pattern for scan angle $-30°$ without mutual coupling (theory-*dashed line*) and measured with mutual coupling (*solid line*)

where $AF_{\text{array}}(\theta, \phi)$=antenna factor related to the linear or planar array. This well-known [15, 21] multiplication formula determines the array radiation pattern as a product of the element radiation pattern and the array factor. Strictly speaking, this formula is valid for infinite equally spaced array. For an ideally matched array with the element gain, which is given by the ratio (2.34), the maximum array gain toward the scanned angles θ_0 and ϕ_0 is given by

$$G_{\text{array}}^{\max}(\theta_0) = \frac{4 \cdot \pi \cdot d_x \cdot d_y}{\lambda^2} \cdot N \cdot \cos\theta_0 \tag{2.45}$$

This is the fundamental upper gain limit for the antenna array with a large number of elements when majority of elements sees the same “neighbors” [36]. Due to the reflection losses $\Gamma(\theta, \phi)$, the realized large array gain when scanning the beam toward the angle θ_0 is given [35]

$$G_{\text{array}}(\theta_0, \phi_0) = \frac{4 \cdot \pi \cdot d_x \cdot d_y}{\lambda^2} \cdot N \cdot \left(1 - |\Gamma(\theta_0, \phi_0)|^2\right) \cdot \cos\theta_0 \tag{2.46}$$

Formula (2.46) assumes identical reflections at each antenna element and does not take into consideration active losses in the feeding network and antenna elements. The gain of a fully exited linear array for $\theta = \theta_0$ and non-identical elements is [35]

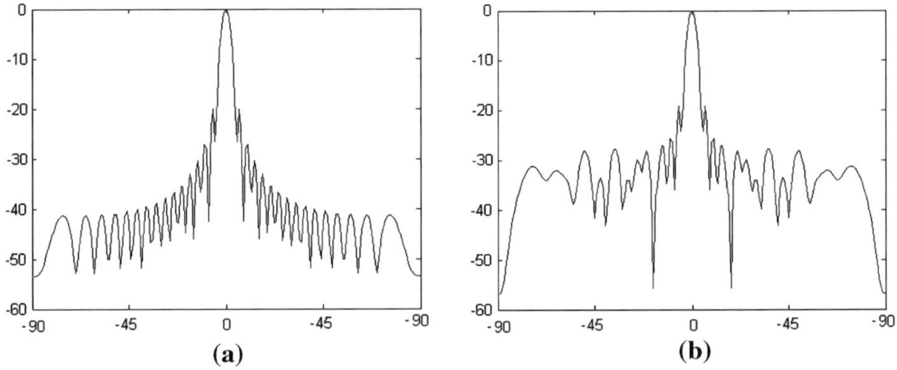

Fig. 2.14 Simulated radiation patterns for 32 element array with ideal element radiation pattern (cos θ): **a** Reference pattern with feeding network providing −28 dB sidelobes; **b** Pattern with phase and amplitude errors

$$G_{\text{array}}(\theta_0) = G_{\text{isolated}}(\theta_0) \cdot \left[N - \frac{\left| \sum_{n=1}^{N} \Gamma_n(\theta_0) \right|^2}{N} \right] \tag{2.47}$$

Mutual coupling is especially pertinent problem [37, 44] when the number of antenna elements is small, often precluding the use of conventional beam synthesis techniques (Chebyshev, Taylor, Bayliss, etc.). Inter-element coupling usually manifests by increased main beamwidth, filling nulls, and raised sidelobes levels. Small arrays are the main candidates for compact automotive applications, and, therefore, estimation of the array radiation pattern with a small element number is very important. Figure 2.13 shows simulation and measurement results for the linear array of X-band rectangular waveguides in the ground plane [37] with mutual coupling between antenna elements. Each element was in turn a column of 8 rectangular waveguides in a common H-plane, combined via a fixed 1:8 power divider.

The array axis, thus, was parallel to the E-plane and in this plane, the element spacing $d = 1.25$ cm $= 0.517\ \lambda$. The isolated element pattern corresponds to a normalized uniform aperture distribution

$$F_{\text{elemnt}}^{\text{isolated}}(\theta) = \frac{\sin\left(\frac{k \cdot l \cdot \sin\theta}{2}\right)}{\frac{k \cdot l \cdot \sin\theta}{2}} \tag{2.48}$$

where $l = 10.2$ mm $= 0.417\ \lambda$. Examples of the element patterns for a central and an edge element are shown in Fig. 2.13a. Figure 2.13b demonstrates synthesized 30 dB Chebyshev patterns obtained with and without the mutual coupling. The radiation pattern, calculated according to the multiplication formula (2.44) is presented by a dashed line, where the radiation pattern of the single element is simulated by formula (2.48). A solid line curve is the array pattern in the E-plane with coupling effect. Figure 2.13c shows two radiation patterns for the scanning angle equal $-30°$: theory calculation without mutual coupling and measurement result obtained with coupling effect, which leads to the increasing of the sidelobe by approximately 10 dB.

Similar distortions of the array radiation pattern occur for the amplitude and phase errors due to the array element tolerances. Let us assume that the amplitudes and phases of the array elements have quasi-random errors due to the manufacturing process, and the antenna array consists of a large number of the elements. In such a situation, errors lead to: directivity (gain) reduction; raising of the average sidelobe level and of some individual sidelobes; errors in the main beam pointing and perhaps shape changing. The ratio of the gain obtained with errors to the gain without amplitude and phase distortions and an approximate average sidelobe level for multi-element array can be estimated as

$$\text{Gain loss} \approx e^{-\sigma^2} \approx \frac{1}{1+\sigma^2}, \quad \text{Average Lidelobe Level} \approx \frac{\sigma^2}{N} \tag{2.49}$$

where $\sigma^2 = \sigma_a^2 + \sigma_\phi^2$, σ_a and σ_ϕ amplitude and phase standard deviations, respectively.

For a uniform square planar array of N elements per side, the standard deviation of the beam pointing error, in terms of the 3 dB beamwidth is [15]

$$\sigma_\theta \approx \frac{0.66 \cdot \sigma}{N} \tag{2.50}$$

Figure 2.14 illustrates two radiation patterns for a 32 element linear array with ideal elements (element pattern is $\cos\theta$): the first one is the reference and it is obtained with a feeding network providing -28 dB sidelobes, while the second radiation pattern is a quasi-random realization of the radiation pattern with phase and amplitude errors ($\sigma_a = 0.1$, $\sigma_\phi = 15°$).

According to the expressions (2.48) and (2.49) expected gain losses are about -0.2 dB, and the average sidelobe level is equal to -28 dB. It is necessary to note that formulas (2.49) and (2.50) present average estimations while different quasi-random realizations of amplitude and phase errors lead to quasi-random radiation pattern parameters.

2.7 Digital Beamforming

In digital beamforming (DBF), the operations of phase shifting, amplitude tapering for each element, and forming of the radiation pattern are produced digitally. From a technological point of view DBF is applicable for both the transmitting and receiving modes. However, for the transmitting mode, beam shape control and pattern nulling are not so critical. For the receiving mode, the main advantages of DBF are as follows:

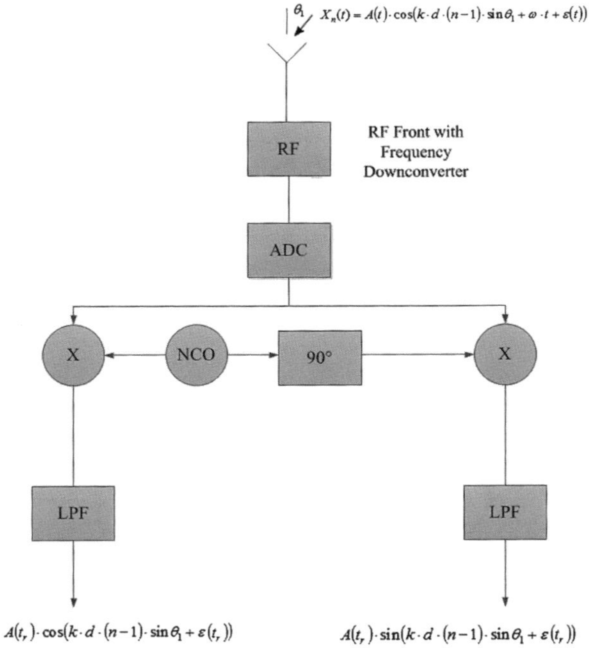

Fig. 2.15 Simplified block diagram of digital array element, produces two quadrature components

a) Multibeam operation which aims to scan wide angle sector with a few narrow beams simultaneously
b) Flexible, accurate, and adaptive pattern shaping, including nulling toward the angle direction of the interference sources and synthesizing of the antenna pattern with extremely low sidelobe level
c) High resolution techniques, which aim to resolve sources with closer angular spacing than one beamwidth
d) Array element radiation pattern correction, which is important for small antenna arrays due to the mutual coupling effects

In this section, we show how to implement basic digital radiation pattern and estimate its main parameters.

In digital antenna array (DAA) system [45], the received signals are digitized at the element level. Since microwave frequency is too high to be directly digitized, each DBF receiver shifts the RF signal frequency down to the low intermediate carrier frequency, and then the low intermediate frequency signal is digitized with an analog to digital (A/D) converter. After the antenna signal has been successfully sampled in the digital domain, the signal needs to be processed by a digital down converter (DDC). DDC multiplies the digital signal with a sinusoidal signal and a 90° phase shifted version of the sinusoidal signal, both are generated by a local numerical controlled oscillator (NCO). Finally, both quadrature digital signals are filtered to suppress the intermediate frequency, creating two zero baseband quadrature signals. Figure 1.8d shows a simplified block diagram of the digital antenna array. Each receiving element of this array shown in Fig. 2.15 has RF front-end with a low noise amplifier and a frequency down convertor, which shifts the signal frequency band to baseband. Analog to digital convertor (ADC), digitizes the baseband signal, and the circuit including numerically controlled oscillator (NCO) produces two quadrature components of the digitized signal that pass through low pass filters (LPF). Let us assume that a plane wave from angle direction θ_1 arrives at the element number n. This RF signal with a carrier frequency ω is modulated with slow time variable amplitude $A(t)$ and phase $\varepsilon(t)$. The circuit shown in Fig. 2.15 produces two quadrature digitized samples at a discrete time t_r; $I_n(t_r) = A(t_r) \cdot \cos(k \cdot d \cdot (n-1) \cdot \sin\theta_1 + \varepsilon(t_r))$ and $Q_n(t_r) = A(t_r) \cdot \sin(k \cdot d \cdot (n-1) \cdot \sin\theta_1 + \varepsilon(t_r))$.

This information, taken for all array elements, determines the amplitude and phase distribution across the array elements due to the wave impinging on the array from the angle direction θ_1. The digital beam processor forms an antenna array radiation pattern similar to the radiation pattern obtained with the electronically controlled phase array. Let us present the output digital signal in complex form (instead of two quadrature signals)

$$S_n(t_r, \theta_1) = I_n(t_r) + j \cdot Q_n(t_r) = A(t_r) \cdot e^{j \cdot k \cdot d \cdot (n-1) \cdot \sin\theta_1 + j \cdot \varepsilon(t_r)} \tag{2.51}$$

Without loss of generality we assume below that the incoming signal is a plane nonmodulated wave, i.e. $A(t_r) \equiv A$; $\varepsilon(t_r) \equiv \varepsilon$; $S_n(t_r, \theta) \equiv S_n(\theta)$. The beamforming processor multiplies the digital codes (2.51) by weighting the coefficients corresponding to some angle direction θ_0. The array factor for (DAA) is determined by:

$$|AF_{\mathrm{DAA}}(\theta_0)| = \left| \sum_{n=1}^{N} S_n(\theta_1) \cdot W_n^*(\theta_0) \right| \tag{2.52}$$

where: $W_n(\theta_0) = \mathrm{Re}W_n(\theta_0) + j \cdot JmW_n(\theta_0)=$ complex weighting coefficients stored in the computer for a number of angles θ_0.

$$W_n = e^{j \cdot k \cdot d \cdot (n-1) \cdot \sin\theta}. \tag{2.53}$$

Using (2.52) and (2.53), the absolute DAA factor is given by

$$|AF_{\mathrm{DAA}}(\theta_0)| = A \cdot \left| \frac{\sin\left(\frac{N}{2} \cdot k \cdot d \cdot (\sin\theta_1 - \sin\theta_0)\right)}{\sin\left(\frac{k \cdot d}{2} \cdot (\sin\theta_1 - \sin\theta_0)\right)} \right| \tag{2.54}$$

The radiation pattern in a digital array can be "scanned" by calculating of the expression (2.54) for different angles θ_0, which are determined by weight coefficients W_n stored in the computer. The process of the radiation pattern calculation is named space beamforming. The beamformer can be implemented by using the Fast Fourier transform (FFT).

This expression is identical to the linear array factor (2.19) obtained for the electronically controllable phase array. If the number of the signal sources with amplitudes is more than one, for example L, and their amplitudes and phases are A_l

and ε_l, respectively, and the angle distance between them is more than the beamwidth of the antenna factor [see expression (2.5)], the array power factor can be expressed as

$$PF_{\text{DAA}} \approx \sum_{l=1}^{L} A_l^2 \cdot \frac{\sin^2\left(\frac{k \cdot d \cdot N \cdot (\sin\theta_l - \sin\theta_0)}{2}\right)}{\sin^2\left(\frac{k \cdot d \cdot (\sin\theta_l - \sin\theta_0)}{2}\right)} \tag{2.55}$$

Therefore, when the sources are separated from each other by the angle exceeding the array beamwidth function, (2.55) exhibits L distinguish peaks, with each peak being located at a point determined by the angle direction of the corresponding source.

To obtain good performance, the phase difference between the quadrature channels should not deviate significantly from 90°. The ADC is a critical element since it limits the system bandwidth and dynamic range. The output of an ADC has 2^R levels, where R = number of converter ADC bits. For example, an ADC with 10-bit output represent 1,024 (2^{10}) signal levels. That means that over the dynamic range from 0 % to 100 %, there will be 1,024 output binary numbers. The bit that represents the smallest voltage change is called the least significant bit. The number of effective bits is limited by the noise floor of the system. The maximum signal to noise ratio SNR and the number of effective bits are related as in [46]

$$\text{SNR} = \text{Number of Effective Bits} \cdot 6.02 + 1.76 \tag{2.56}$$

SNR is expressed in dB scale.

The other important ADC parameters are bandwidth and sampling rate. Bandwidth describes the frequency range in which the input signal can pass through the analog section of the ADC with minimal amplitude loss. It is recommended that the bandwidth of the digitizer be 3 to 5 times the highest frequency component of interest in the measured signal with minimal amplitude error [47]. The theoretical amplitude error of a measured signal is calculated from the ratio of the digitizer's bandwidth in relation to the input signal frequency f_{input}

$$\text{Error}(\%) = \left(1 - \frac{\text{ratio}}{\sqrt{1 + \text{ratio}}}\right) \cdot 100 \tag{2.57}$$

where ratio $= \frac{\text{banwidth}}{f_{\text{input}}}$.

The sampling rate has to be greater than the theoretical Nyquist rate. The Nyquist theorem states that to avoid aliasing, a signal must be sampled at a rate greater than twice the highest frequency component of the signal to reconstruct the waveform accurately. However, to digitize the incoming signal, the digitizer's real-time sample rate should be at least three to four times the digitizer's bandwidth [48].

In addition, it is necessary to note that for proper operation of digital beamformer processor, the ADCs of all array elements have to convert analog signals to digital codes synchronically, otherwise the digital outputs will produce amplitude and phase errors and eventually distort the computed antenna array factor [49]. The calibration of an ADC [50] is a key initial step in the successful application of algorithms reliant on spatial relationships across a multiple-antenna system.

References

1. Thiel A et al (2010) In situ vehicular antenna integration and design aspects for vehicle to vehicle communications, antennas and propagation (EuCap). In: Proceedings in the Fourth European Conference on 12–16 April 2010
2. Schack M et al (2010) Analysis of channel parameters for different antenna configurations in vehicular environments. Vehicular Technology Conference, IEEE, 2010
3. Sugiura S, Iizuka H (2007) Reactively steered ring antenna array for automotive application. IEEE Trans Antennas Propag, V 55(7):1902–1908
4. Song H et al (2002) Diversity antenna studies at 2.3 GHz for automotive applications, and MEMS-based antenna diversity switching circuit. In: Proceedings VTC 2002-Fall. 2002 IEEE 56thVehicular Technology Conference, 2002
5. Nakano H et al (2010) Array antenna composed of bent four leaf elements. 3rd european conference on antennas and propagation EuCAP 2009
6. Rabinovich V et al (2010) Automotive antenna design and applications. CRC Press, Boca Raton
7. Nakanishi T et al (2006) Multiple-loop array antenna with switched beam for short range radars. Vehicular Technology Conference, 2006 VTC-2006 Fall, 2006 64th, 25–28 Sept 2006
8. Chen Z, Luk K (2009) Antennas for base stations in wireless communications. McGraw-Hill, New York
9. Choi B et al (2009) Circularly polarized H-shaped microstrip array antenna with T-slot for DSRC system roadside equipment. Microwave Opt Technol Lett, 51(6):1545–1548
10. Song C et al (2005) Compact low cost dual polarized adaptive planar phased Array for WLAN, IEEE Trans Antennas Propag 53(8):2406–2415
11. Mase K et al (2008) Performance evaluation of roadside to vehicle communication system using narrow antenna beam switching based on traffic flow model. GLOBECOM, Workshops
12. Song H et al (2002) Diversity antenna studies at 2.3 GHz for automotive applications, and mems-based antenna diversity switching circuit. In: Proceedings VTS 2002-Fall, IEEE 56th Vehicular Technology Conference pp 1096–1099
13. Khaleghi A et al (2005) Diversity techniques with dipole antennas in indoor multipath propagation. In: IEEE 16th international symposium on personal, indoor and mobile radio communication pp 669–673
14. Monzingo RA, Miller TW (2005) Introduction to adaptive arrays, SciTech Publishing
15. Robert C Hansen (2009) Phased array antennas, 2nd edn. Wiley, New York
16. Shen W et al (2007) A compact high performance linear series-fed printed circuit antenna array. In: ICMMT-07, International conference on Microwave and Millimeter Wave Technology
17. Lin C et al (2010) A compact linearly polarized antenna array with low sidelobe. In: International Conference on Microwave and Millimeter Wave Technology (ICMMT) pp 384–387
18. Gruszczynski S et al (2006) Reduced sidelobe four-beam n element antenna arrays by 4N butler matrixes. IEEE Antennas Wirel Propag Lett 2:430–434
19. Hancen RC (1985) Aperture efficiency of villeneuve $\bar{n}$ arrays. IEEE Antennas Propag 33(6):666–669
20. Randy L (2010) Haupt antenna arrays: a computational approach. Wiley, New York
21. Balanis C (1997) Antenna theory analysis and design, 2nd edn. Wiley, New York

22. MacPhie R (2007) A mills multiplicative array with the power pattern of a conventional planar array. In: Antennas and propagation society international symposium, IEEE, pp 5961–5964
23. Aumann HM (2010) A pattern synthesis technique for multiplicative arrays. In: PIERS Proceedings, Cambridge, USA pp 864–867
24. Mills B, et al (1985) A high resolution radio telescope for use at 3.5 m. In: Proceedings of the IRE, pp 67–84
25. Forman BJ (1972) Directivity characteristics of scannable planar arrays. IEEE Trans Antennas Propag 20(3):245–252
26. Lee M et al (2000) Evaluation of directivity for planar antenna arrays. IEEE Antennas Propag Mag 42(3):64–67
27. Ioannides P, Balanis C (2005) Uniform circular and rectangular arrays for adaptive beamforming applications. IEEE Antennas Wirel Propag Lett 4:351–354
28. Rabinovich V (1975) Statistics of the array with digital phase shifters. J Commun Technol Electron (in Russia) 4:708–714
29. Smith M, Guo Y (1983) A comparison of methods for randomizing phase quantization errors in phased arrays. IEEE Trans Antennas Propag 31(6):821–828
30. Taheri S, Farzaneh F (2006) New methods of reducing the phase quantization error effects on beam pointing and parasitic side lobe level of the phased array antennas. Microwave Conference APMC 2006. Asia Pacific, pp 2114–2117
31. Guo Y et al (2003) Comparision of random phasing methods for reducing beam pointing errors in phased array. IEEE Trans Antennas Propag 51(4):782–787
32. Aronov FA (1966) New method of phasing for phased array using digital phase shifters. Radio Eng Electron Phys 11:1035–1040
33. Clenet M, Morin G (2003) Visualization of radiation-pattern characteristics of phased arrays using digital phase shifters. IEEE Antennas Propag Mag 45(2):20–35
34. Maloratsky L (2010) Electrically tunable switched line diode phase shifters. High Frequency Electronics 9(4):part1, 9(5):part2
35. Pozar D (1994) The active element pattern. IEEE Trans Antennas Propag 42(8):1176–1178
36. Hannan P (1964) The element gain paradox for a phased array antenna. IEEE Trans Antennas Propag 12(4):423–433
37. Steyskal H, Herd J (1990) Mutual coupling compensation in small array antennas. IEEE Trans Propag 38(12):1971–1975
38. Pozar D (2003) A relation between the active input impedance and the active element antenna pattern of phased array. IEEE Trans Antennas Propag 51(9):2486–2489
39. Fenn A (1985) Theoretical and experimental study of monopole phased array antennas. IEEE Trans Antennas Propag 33(10):1118–1126
40. Fenn A, Willwerth F (1984) Mutual coupling in monopole phased arrays. In: Antennas and Propagation Society Symposium 22:875–878
41. Pozar D (1986) Finite phased arrays of rectangular microstrip patches. IEEE Trans Antennas Propag 34(5):658–665
42. Jedlicka R et al (1981) Measured mutual coupling between microstrip antennas. IEEE Trans Antennas Propag 29(1):147–149
43. Pozar D (1981) Input impedance and mutual coupling of rectangular microstrip antennas. IEEE Trans Antennas Propag 30(6):1191–1196
44. Borowiec R et al (2002) Compensation of mutual coupling in small antenna arrays. In: International conference microwave, radars and wireless communication pp 894–897
45. Steyskal H (1987) Digital beamforming antennas: An introduction. Microwave J. 30(1):107–124
46. Tran T (2010) High speed DSP and analog system design, Springer Chapter 7
47. Bandwidth, Sample Rate, and Nyquist Theorem. NI Developer Zone - National Instrument www.ni.com/white-paper/2709/en
48. Top 10, Things to Consider When Selecting a Digitizer/Oscilloscope. NI Developer Zone - National Instrument www.ni.com/white-paper/4333/en

49. Bratchikov A, Dobycina E (2009) Digital antenna array calibration. In: 19th International crimean conference microwave and telecommunication technology, pp 401–402
50. Tyler N, Allen B, Aghvami H (2004) Adaptive antennas: the calibration problem. Commun Mag IEEE 42(12) pp 114–122

Chapter 3
Smart Beamforming: Main Adaptive Techniques

3.1 Adaptive Systems

3.1.1 Introduction

The adaptive arrays are expected to dramatically improve the performance of wireless communications systems because they have the potential to expand coverage, increase the capacity, and improve the signal quality. They utilize sophisticated signal processing algorithms to form beam minima toward the interfering signals, continuous beam steering, to distinguish between desired signals, and to calculate their directions of arrival. An adaptive array was first introduced by Van Atta in 1959 [1] to describe a self-phased array, which reflected all incident signals back in the direction of arrival by using phase conjugation circuits. Later, Applebaum developed an analog sidelobe canceler for automatic interference nulling [2, 3], which maximizes the signal-to-noise ratio. This technique is named the Howells–Applebaum algorithm. Another popular algorithm, named as Least Mean Squares (LMS), developed by Widrow [4], minimizes the mean square error (MSE) between the input and the reference signal. In the early years Howells–Applebaum and Widrow antennas were utilized exclusively in radar systems, however, they have now become standard devices in commercial wireless communication where it is necessary to improve quality in the multipath and noise environment. The excellent reference book [5] written by Monzingo, Haupt, and Miller describes different algorithms employed in the adaptive array technology. Adaptive antennas can be implemented as the analog or digital devices. Generally, both of them regulate amplitudes and phases (weight coefficients), sometimes only amplitudes or only phases of the array elements. Adaptive algorithm provides the shape of radiation pattern according to the certain criteria, for example, with nulls in the directions of the incoming interferences sources or MSE between a signal arriving from the desired angle direction and an output array signal. The analog design of the fully adaptive system assumes that each

V. Rabinovich and N. Alexandrov, *Antenna Arrays and Automotive Applications*,
DOI: 10.1007/978-1-4614-1074-4_3,

antenna array element has an analog receiver that controls amplitude and phase of the signal passing through the element channel. In the digital beamforming system, the output signal from each element is digitized with the analog to digital (A/D) convertor and digital signals are processed in the adaptive digital beamformer. The digital beamformer is a very flexible device that adapts weighting coefficients according to a number of different algorithms. However, the frequency of the radio signals is too high to be directly digitized, and, therefore, digital beamforming receivers use analog "RF translators" to shift the signal frequency down before the ADC, as shown in Fig. 2.15. The ADC have to provide high dynamic range and speed of operation (depending on the signal bandwidth).

Partially, adaptive systems [6, 7] mainly use high gain antenna or the array without adaptive elements and an additional low gain adaptively controlled array which serves to suppress interference sources while keeping the main beam without significant degradation.

Adaptive technique, based on perturbation of element phase only, is very attractive in a phased array because the required control is available at no extra cost. Adaptive phase control methods have been extensively addressed in the literature [8–16]. For example, Steyskal [9] developed a linearization method to place nulls at specified directions using small phase perturbation. Baird and Rassweiler [10] have shown that the optimal phase distribution across an array with phase-only nulls placed at known directions can be exactly represented as the phase of a weighted sum of partial beams pointed at the desired signal and interferences. Such representation significantly reduces dimensionality of the nonlinear problem when the number of interferences is small. A number of authors [11–15] investigated the phase-based algorithm using minimum output array power criteria. Some of these algorithms adjust phase setting, using the random phase perturbation, to provide the minimum output power. Some of them [12] form an approximate numerical gradient and use a steepest descent method to find minimum power. In order to avoid a degradation of the main beam, Haupt [11] used only the least significant bits of the digital phase shifters. One of the high effective beam space algorithms assumes that the location of the interference is known. This algorithm forms a cancellation beam in the direction of the interference. The level of the cancellation beam is adjusted to cancel the sidelobes in the corresponding direction. This method is fast, however, it requires the knowledge of the interference source angle location. Davis [16] developed a phase-only gradient-based adaptive algorithm that estimates the search direction using cross-correlating binary perturbation sequences.

3.1.2 *Maximum Signal-to-Noise Ratio: Applebaum Loop*

3.1.2.1 Full Adaptive Configuration

Consider an N-element adaptive array. Elements of the array are assumed isotropic and separated by a half wavelength. A typical block scheme of a full adaptive array with variable element weights w_n ($n = 1, 2, 3 \ldots N$) is shown in Fig. 1.8a. A few interference signals impinge on the array elements from different angle directions, and the receivers employed for realization of adaptive process are noisy. Antenna is assumed to be a steered beam adaptive array; noise signal received by nth array element is characterized by complex envelope $x_n(t)$

$$x_n(t) = \sum_{r=1}^{R} A_r(t) \cdot e^{j \cdot k \cdot d \cdot (n-1) \cdot \sin \theta_r + j \cdot \varepsilon_r(t)} \tag{3.1}$$

where r = number of the interference source ($r = 1, 2 \ldots R$), $A_r(t)$ = amplitude of the signal with the index r, $\varepsilon_r(t)$ = phase of the signal with the index r, and θ_r = angle direction of the corresponding external noise source. Complex signal Eq. (3.1) received by each antenna element is multiplied by a complex weight $w_n(t)$ to produce combined array output $y(t) = \sum_{n=1}^{N} w_n(t) \cdot x_n(t)$. The average noise power in the kth channel (before weight multiplying) is denoted by μ_{kk}, and the covariance of $x_k(t)$ is given by

$$\mu_{kl} = \overline{(x_k(t)^* \cdot x_l(t))} \tag{3.2}$$

The upper line indicates math expectation by time t. Asterisk (*) denotes the complex conjugate. Optimal Applebaum maximum signal-to-noise ratio (MSN) solution [2] for steady-state weight coefficients can be expressed as

$$\mathbf{W}_{\text{opt}}^{\text{ss}} = \varepsilon \cdot \mathbf{M}^{-1} \cdot \mathbf{S}^* \tag{3.3}$$

where $\mathbf{W}_{\text{opt}}^{\text{ss}}$ = column vector, which consists of N weight components w_n^{ss}; $\mathbf{M} = [\mu_{kl}]$ = covariance matrix from components Eq. (3.2); $\mathbf{M}^{-1}$ = inverse of matrix M, S = column vector with the elements $S_n = e^{j \cdot k \cdot d \cdot (n-1) \cdot \sin \theta_0}$ (θ_0 is an angle of desired signal); ε = arbitrary constant.

The steady-state array factor (after adaptation) at the angle direction of the incoming interference signal is given by

$$\text{AF}_{\text{adlinear}}(\theta_0, \theta_1) = \left(\frac{p_0}{p_0 + N \cdot p_1} \right) \cdot \text{AF}_{\text{linear}}(\theta_0, \theta_1) \tag{3.4}$$

where value $\text{AF}_{\text{linear}}(\theta_0, \theta_1)$ for uniform amplitude and phase distribution is determined by the formula (2.19), p_0 = average thermal receiver noise power level of each array element, p_1 = average (over time t) interference source power.

Hence, the adaptive control reduces the gain in the direction of noise source by the factor

$$Q = \frac{1}{1 + \frac{N \cdot p_1}{p_0}} \tag{3.5}$$

Ratio Eq. (3.5) shows that level of the interfering source after adaptation cycle reduces with the increase of the element numbers and the power ratio between the jammer and the thermal element power noise level.

The Applebaum system provides interfering signal suppression effectively when the number of jammers is less than the number of array elements, and when the angle direction of the jammer does not coincide with the main beam angle direction. Figure 3.1a shows the sketch of the radiation pattern after adaptation. Resultant pattern is the difference between the quiescent array and cancellation pattern, which is produced by the weight coefficients w_n^{ss}.

The preceeding formulas present the steady-state solution for adaptive coefficients. In the real environment, when interference signal impinges on the antenna array, the processor dynamically adapts the weighting coefficients to steady-state values Eq. (3.3). Figure 3.1b shows a practical electronic circuit loop built at each antenna array element according to the Howells–Applebaum algorithm. Adaptive processing operation of the loop is given by the following equation

$$\tau \cdot \frac{\overline{w_i(t)}}{dt} + \overline{w_i(t)} = G\left\{ S_i^* - \sum_{n=1}^{N} \overline{w_n(t)} \cdot \overline{x_i^*(t) \cdot x_n(t)} \right\} \tag{3.6}$$

where τ = time constant of the integrated circuit, G = amplifier gain, $S_n = e^{j \cdot k \cdot d \cdot (n-1) \cdot \sin \theta_0}$ = complex coefficient, which specifies the desired signal arrival angle θ_0.

Simulation results of the full adaptive array consisting of eight, four, and two isotropic elements are presented in Figs. 3.2, 3.3, and 3.4. All simulation radiation patterns demonstrate steady-state regime for signal-to-noise ratio equal to 30 dB. Arrows in figures show the angle positions of the interference sources. Graphs in Fig.3.2 show the radiation patterns of the antenna array, which consists of eight isotropic antenna elements. Each antenna element provides amplitude and phase adaptive regulations of the passing through the antenna elements signal. Figure 3.2a presents the quiescent pattern before adaptation. Figure 3.2b–d demonstrates the radiation pattern when one, two, and three interference sources impinge on the adaptive array system. As we can see, the signal levels toward the interference sources are significantly suppressed for all three scenarios. Because the array with a small element number is our prime interest we also present simulation results for the antenna with four and two adaptive elements. Figure 3.3 shows the results related to four array elements and Fig. 3.4 presents the adaptive radiation pattern for two array elements.

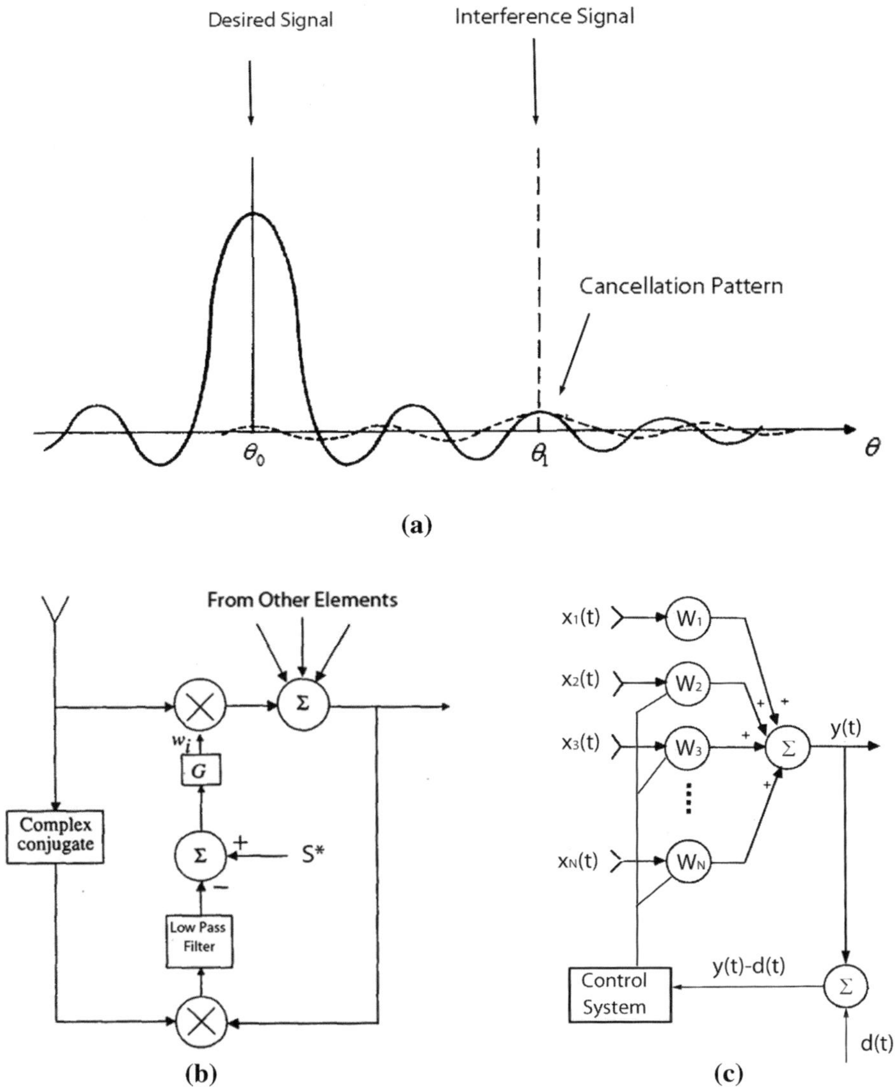

Fig. 3.1 **a** Radiation pattern concept of the Applebaum array; **b** Howells–Applebaum adaptive loop circuit; **c** Adaptive array based on LMS algorithm

3.1.2.2 Partially Adaptive Topology with Amplitude and Phase Regulation

Figure 3.5 demonstrates simulation results for the cost-effective partially adaptive array. Such system includes the main array or one antenna with non-adaptive weight coefficients and a few adaptive elements.

Figure 3.5a–c is related to the array with six regular elements (with no adaptive loop), and two edge elements include the adaptive weights. Figure 3.5d and e

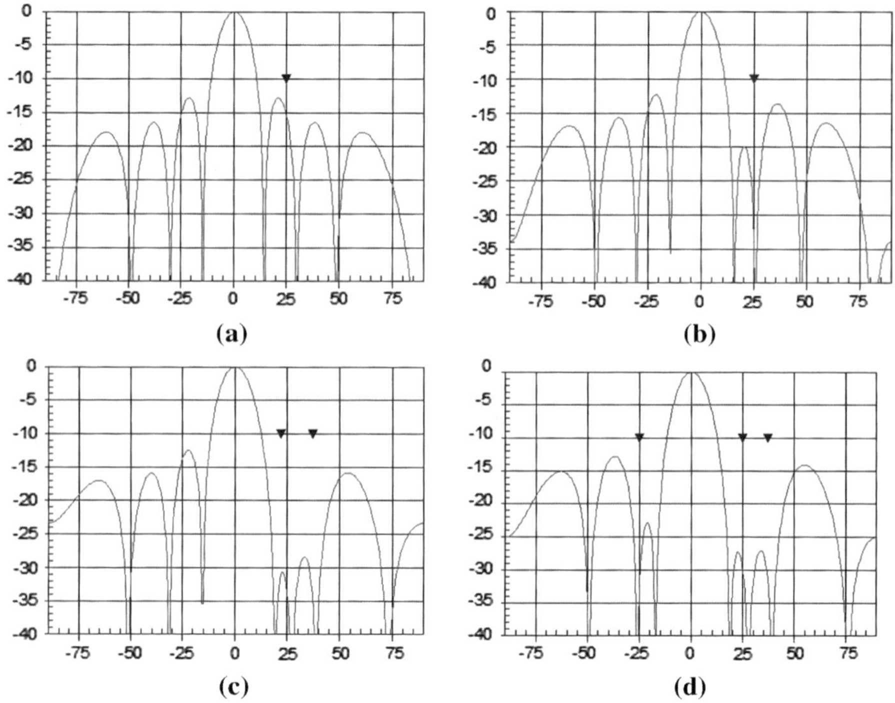

Fig. 3.2 Steady-state radiation pattern of full adaptive eight-element array: **a** Radiation pattern before adaptation; **b** Radiation pattern after adaptation, one interference source; **c** Two interference sources; **d** Three interferences

represents the radiation pattern of the array with one adaptive element and seven non-adaptive elements. It is seen that the partially adaptive array with only one adaptive element shows reasonable results when suppressing one interference source.

3.1.3 Least Mean Square Algorithm

The least mean squares (LMS) algorithm was developed by Widrow [4] and can be explained as follows (Fig. 3.1c). Assume that the signal $x_i(t)$, impinging on the array element number i, consists of the desired signal, a few interfering signals, and noise. The purpose of this algorithm is to choose weight coefficients that provide MSE between the desired signal d*(t)* and resultant output weighted signal $y(t) = \sum_{i=1}^{N} w_i(t) \cdot x_i(t)$.

$$\overline{|\varepsilon(t)|^2} = \overline{|d(t) - y(t)|^2} = \min \tag{3.7}$$

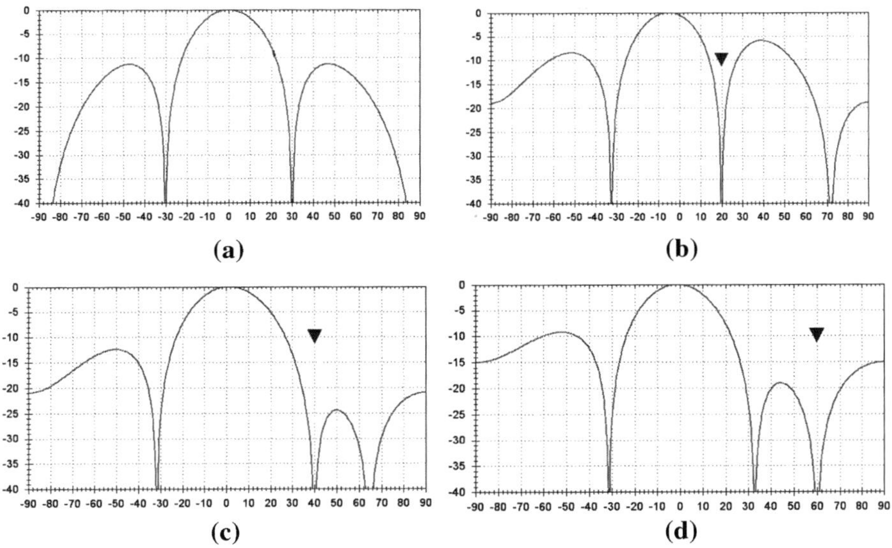

Fig. 3.3 Steady-state radiation pattern of full adaptive four-element array versus the interference source angle position: **a** Radiation pattern before adaptation; **b** Source angle 20°; **c** Source angle 40°; **d** Source angle 60°

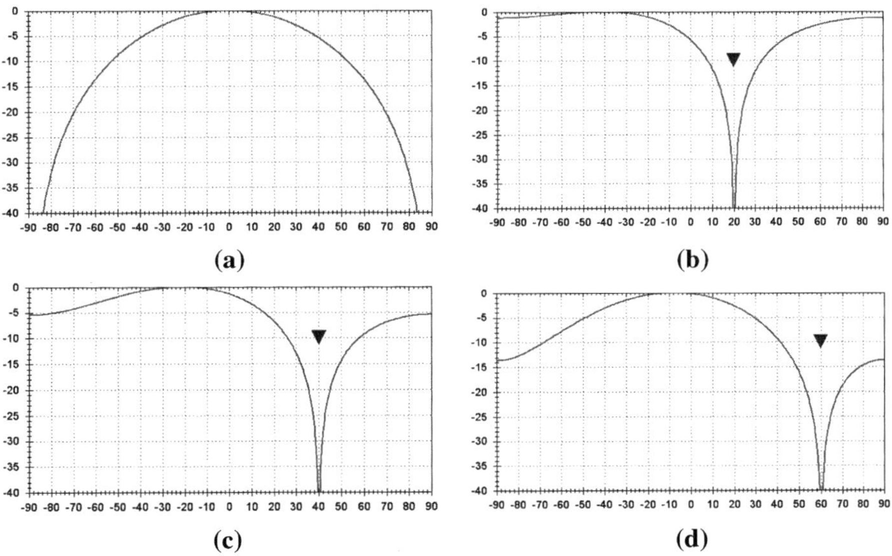

Fig. 3.4 Steady-state radiation pattern of full adaptive two-element array versus the interference source angle position: **a** Radiation pattern before adaptation; **b** Source angle 20°; **c** Source angle 40°; **d** Source angle 60°

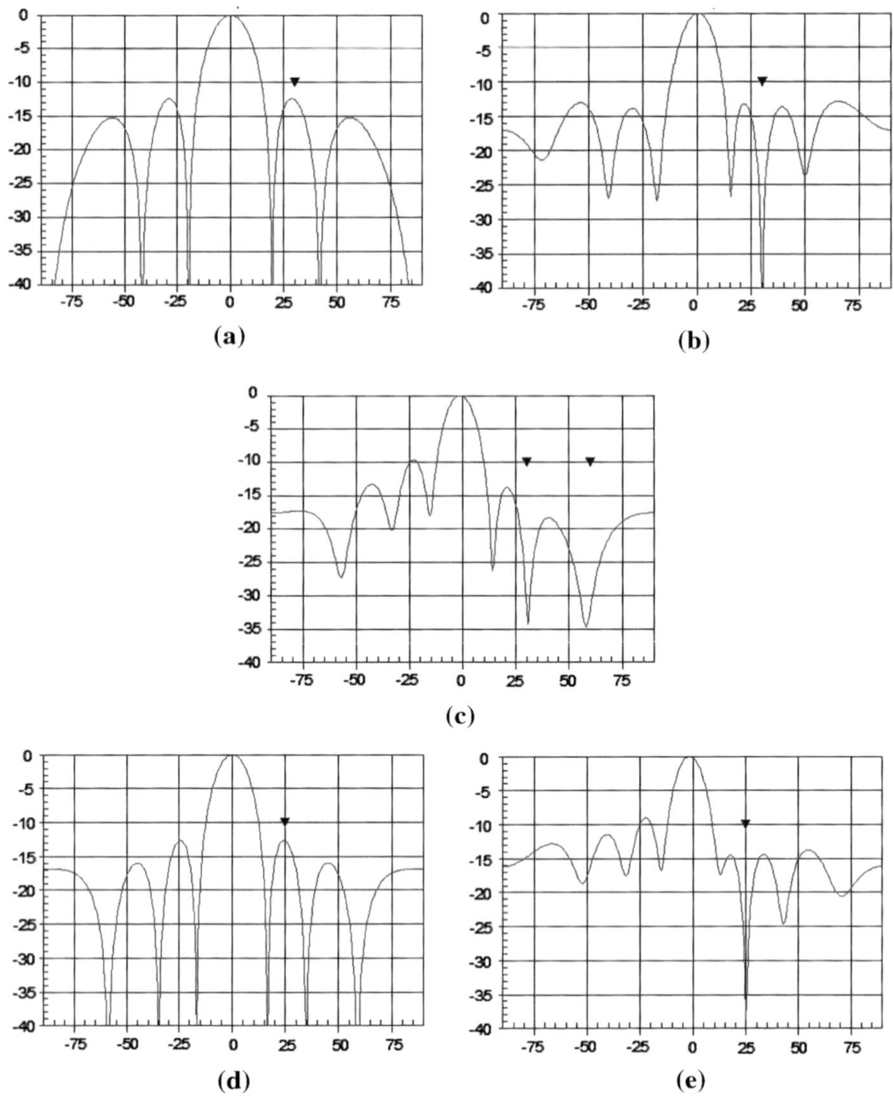

Fig. 3.5 Simulation Results of the partially adaptive array: **a** to **c** Six non-adaptive and two adaptive elements: **a** Radiation pattern before adaptation, six elements; **b** One interference, 30°; **c** Two interferences, 30° and 60°; **d** Radiation pattern before adaptation, seven elements; **e** Radiation pattern after adaptation, seven non-adaptive and one adaptive elements, one interference, angle 25°

Time averaging in Eq. (3.7) is symbolically denoted by an upper overall bar. The optimal solution for the adaptive weight coefficients are given

$$\mathbf{W_{LMS}} = \mu_0 \cdot \mathbf{R_{xx}^{-1} r_{xd}} \tag{3.8}$$

where the covariance matrix $\mathbf{R_{xx}}$is given by

$$\mathbf{R_{xx}} = \begin{bmatrix} \overline{x_1(t) \cdot x_1^*(t)} & \dots & \overline{x_1(t) \cdot x_N^*(t)} \\ \overline{x_2(t) \cdot x_1^*(t)} & \dots & \overline{x_2(t) \cdot x_N^*(t)} \\ \overline{x_N(t) \cdot x_1^*(t)} & \dots & \overline{x_N(t) \cdot x_N^*(t)} \end{bmatrix} \tag{3.9}$$

and the vector $\mathbf{r_{xd}}$ is expressed as

$$\mathbf{r_{xd}} = \begin{bmatrix} \overline{x_1(t) \cdot d^*(t)} \\ \overline{\dots\dots} \\ \overline{x_N(t) \cdot d^*(t)} \end{bmatrix} \tag{3.10}$$

and μ_0 is a constant value.

Gradient solution for the weighting coefficients $w_n(t_q)$in digital form is expressed as

$$\overline{w_n(t_{q+1})} = \overline{w_n(t_q)} + \mu_1 \cdot \overline{\varepsilon^*(t_q) \cdot x_n(t_q)} \tag{3.11}$$

where $t_q =$ discrete time, which can be expressed as $t_q = q \cdot \Delta t$. Upper line in formula (3.11) means averaging over fast time fluctuations. The convergence of the algorithm (3.11) is proportional to the step size μ_1. If the step size is too small, convergence is slow. If the convergence is too slow, the adaptive array cannot track the changing signal. If the convergence is to fast (step size $\mu_1$is too large), the adaptive weights oscillate around the optimum values. Stability of convergence is provided [5] when

$$0 \le \mu_1 \le \frac{1}{2 \cdot \lambda_{\max}} \tag{3.12}$$

where $\lambda_{\max} =$ the largest eigenvalue of matrix Eq. (3.9).

3.1.4 Adaptation Based on Phase Variations

Phase shifters are part of the phased array, and, therefore, adaptation using only these electronic elements is very attractive for such types of antennas. As we noted before, a number of reference studies [8–16] have addressed phase only nulling in low-sidelobe antennas. Genetic algorithm for large antenna arrays [11, 12] is one of the most effective technique that demonstrates fast convergence and high nulling level toward the interferences. The algorithm is based on the minimization of the output power by perturbation of a few last bits of the digital phase shifters. Detailed discussion of genetic method using special computer program is beyond the scope of this book; therefore, here we only show some results of simulations that conform the efficiency of this method.

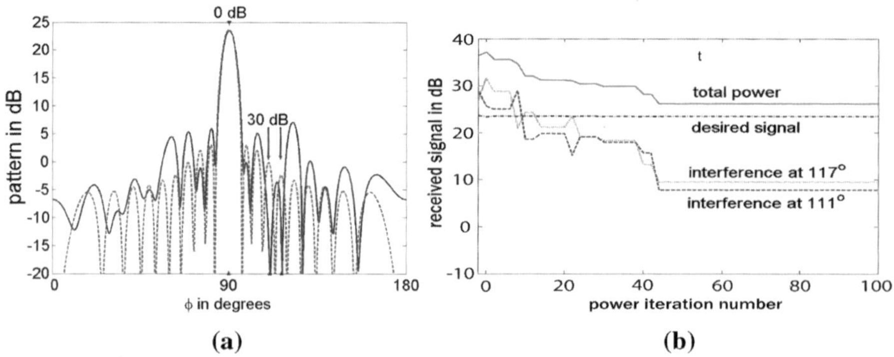

Fig. 3.6 Simulation results for 20 elements of fully adaptive array with phase perturbation: **a** Steady-state radiation pattern, two interference sources at the angles 111° and 117°; **b** Convergence curve of genetic algorithm, [12] © 2006 IEEE

Figure 3.6 presents the simulation results [12] for 20 elements of fully adaptive array with only phase perturbation.

Each antenna element has a six-bit phase shifter, which corresponds to the following phase settings:

Bit	1	2	3	4	5	6
Phase	π	$\pi/2$	$\pi/4$	$\pi/8$	$\pi/16$	$\pi/32$

Only three smallest phase discrete $\pi/32$, $\pi/16$, and $\pi/8$ are used to minimize the output power. It is important to note that these last three bits do not significantly perturb the main beam. Elements of the array are 0.5λ apart. There is 20 dB low sidelobe Taylor amplitude taper. Figure 3.6a presents the steady-state radiation pattern for two 30 dB interference plane waves incident to the array at the angles 111° and 117°. Convergence curve for the algorithm is shown in Fig. 3.6b. The output power decreases monotonically, but the power recieved from the individual interference source varies. At times, the power received by the array from one interference source will go up or down relative to the power received from the second interference source. The output power of the desired signal shows little variation because the main beam remains virtually unperturbed.

3.2 Systems with Multi-Antenna Elements at the Transmitter and Receiver Ends

3.2.1 SISO, MISO, SIMO, and MIMO Design: Brief Introduction

Standard wireless communication system includes the following components: a transmitter with a transmitting antenna, propagation channel, and a receiver with a receiving antenna. Various propagation effects, such as reflections from the ground, buildings, and surrounding vehicles (multipath propagation) cause the distortions of the received signal. This means that each antenna element receives direct signal from the transmitting antenna as well as reflecting waves from the surrounding objects. For example, when two radio waves with opposite phases impinge on the antenna, the received radio signal becomes weaker. Sometimes, not only one reflecting wave, but several waves reach the antenna element from different reflection paths. The reception quality is also degraded due to the car body effect. For example, the directionality of an antenna inserted in a car glass window is not omnidirectional. It has a few directionality dips over 360°. Toward the angle "dip", the signal-to-noise ratio of the wireless system can be dramatically reduced.

MIMO systems were developed to improve the quality of reception when the wireless system suffers from the multipath fluctuations and distortions in the antenna's radiation pattern. Multiple antenna systems at the transmitter and receiver sides as well as special transmitting and receiving techniques can significantly enhance the reception quality in reach multipath environment.

A few techniques utilize multiple antennas from the transmitter side. For example, spatial multiplexing (SM) is a technique in which independent and separately encoded data streams are transmitted simultaneously in parallel channels from each element in an array. Using this scheme, the bit stream of data to be transmitted is de-multiplexed into two sub-streams, then modulated and transmitted simultaneously from each transmit array antenna element. At the receiving end, the signal is demodulated, the original sub-streams are combined to yield the original bit stream of data. An alternative method designed for use with multiple transmit antenna elements, known as space time coding (STC), uses redundant copies data stream from the transmitter to the receiver in the hope that some of them are not distorted when propagation through the physical path between the transmitter and the receiver [13]. Therefore, both the antenna array and propagation channel with statistical description in the propagation environment have to be treated together for benefit estimation of multi-element systems.

A few different multi-element antenna system configurations are applicable to increase reception quality in the multipath propagation environment. Figure 3.7 shows several basic diagrams of the MIMO topologies. Each arrow represents the combination of all signals between two antennas that include the direct line of sight (LOS) path and multipath signals created by reflections from the surrounding environment, as shown schematically in Fig. 3.7b.

The general MIMO system presented in Fig. 3.7a includes an array with M transmit elements T_1, $T_2 \ldots T_M$, an array with N receive elements $R_1\, R_2 \ldots R_N$, and the propagation channel with a number of electromagnetic waves between transmit and receive arrays. The SIMO system (Fig. 3.7c) uses single the transmitter and multi-element receiving antenna, the MISO topology presented in Fig. 3.7d uses multi-element transmitter array and the single receiver antenna, and single input single output (SISO) design is shown in Fig. 3.7e. In the SIMO system, benefit in signal to noise ratio can be reached by space, polarization, and pattern diversity techniques, or adaptive beamforming, and beam steering methods.

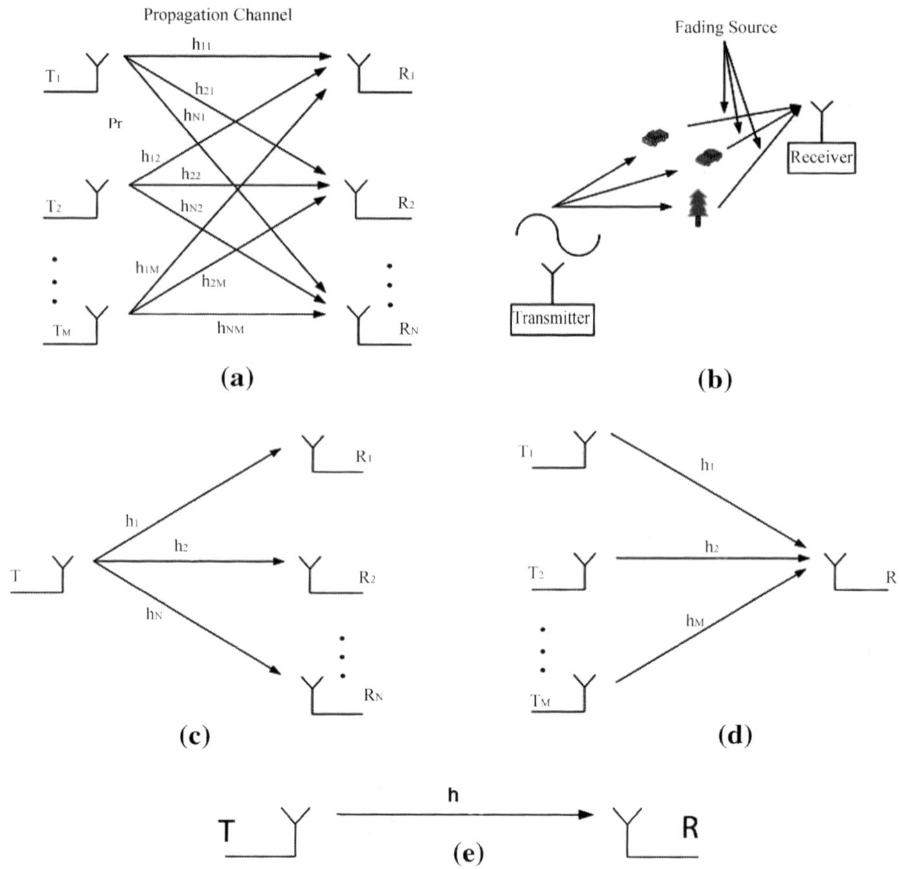

Fig. 3.7 **a** MIMO system; **b** Propagation channel concept; **c** SIMO configuration; **d** MISO design; **e** SISO channel

One of the most important parameters named as the channel capacity was introduced by Shannon in 1958. Shannon channel capacity defines the theoretical maximum information transfer rate in bit per second (bit/s) that can be transferred through the noisy communication channel. The simplest channel includes the wireless communication link between the transmitting and the receiving antennas and the thermal noise caused by electronic devices. In a changing link, such as mobile communication, the propagation channel becomes random variable, and is estimated by statistical parameters for link characterization. As it is shown in numerous papers, the MIMO systems improve the channel capacity in the communication systems.

3.2.2 SISO System and Communication Channel

The simplest scenario shown in Fig. 3.7e includes one transmit antenna, one receive antenna, and a multipath propagation channel characterized by the transfer coefficient h. The transmitting or receiving antennas can be implemented as the single antenna, fixed beam antenna array, or phased array with beam steering. The antenna system has a single output, which is connected with a transmitter from one end and with a receiver from the other end. Let us express the received signal y as

$$y = x \cdot h + n \tag{3.13}$$

The average received power is given by

$$P_{r1} = P_r + P_n = |x|^2 \cdot \overline{|h|^2} + \overline{|n|^2} \tag{3.14}$$

where $P_n = \overline{|n|^2} = N_0$ is a thermal averaged receiver noise power. Typically, thermal noise is a random value: the real and imaginary parts of the noise are statistically independent Gaussian random variables with zero mean and variance σ^2, and statistical expectation equal $P_n = 2 \cdot \sigma^2 = N_0$. Value P_r can be expressed with Friis transmission formula

$$P_r = \frac{P_t \cdot G_t \cdot G_r \cdot \left|F_t^{\mathrm{norm}}(\theta_0, \phi_0)\right|^2 \cdot \left|F_r^{\mathrm{norm}}(\theta_0, \phi_0)\right|^2 \cdot \lambda^2}{(4 \cdot \pi \cdot R_{11})^2} \cdot \overline{|h|^2} \tag{3.15}$$

where G_t and G_r are the gains of transmit and receive antenna, respectively, and $F_t^{\mathrm{norm}}(\theta, \phi)$ and $F_r^{\mathrm{norm}}(\theta, \phi)$are normalized radiation patterns for transmit and receive antennas, R_{11} = distance between transmitting and receiving antennas, λ = wavelength, θ_0 and ϕ_0 angles, which determine the direction from transmitting to receiving antenna. Multiplier $\overline{|h|^2}$ in formula (3.15) is a propagation channel coefficient, which equals to one only for free space propagation. In general, coefficient h is a random value, which reveals the properties of the fading environment (propagation channel) and, therefore, is specified by certain statistical parameters. Comparing value (3.15) and value P_r in relationship (3.14) the ratio $|x|^2$can be written as

$$|x|^2 = \frac{P_t \cdot G_t \cdot G_r \cdot \left|F_t^{\mathrm{norm}}(\theta_0, \varphi_0)\right|^2 \cdot \left|F_r^{\mathrm{norm}}(\theta_0, \varphi_0)\right|^2 \cdot \lambda^2}{(4 \cdot \pi \cdot R_{11})^2} \tag{3.16}$$

So, Eq. (3.16) presents the coefficient, which depends on the transmitting and receiving antenna parameters, the distance between antennas, and wavelength. The signal-to-noise ratio SNR is given by

$$\mathrm{SNR} = \frac{P_r}{P_n} = \frac{|x|^2 \cdot \overline{|h|^2}}{N_0} = \rho \cdot \overline{|h|^2} \tag{3.17}$$

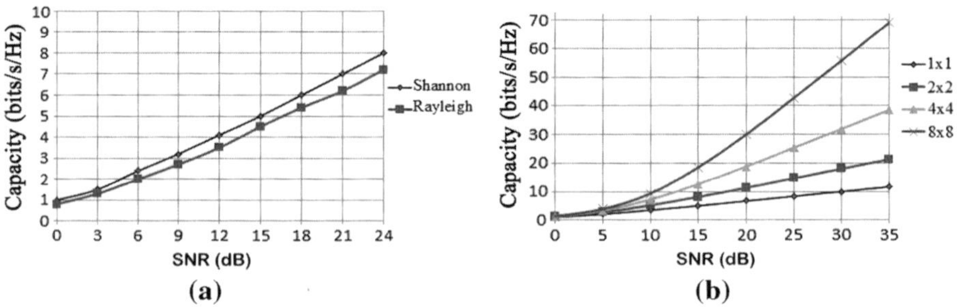

Fig. 3.8 Average capacity versus SNR: **a** SISO channel; **b** MIMO configuration with equal number of the transmitting and receiving elements, additive white Gaussian noise

In this representation, the total SNR is decomposed into two multipliers: first factor ρ is determined by the transmitting power, antenna parameters, free space path loss and thermal receiver noise power. The second factor $\overline{|h|^2}$ characterizes only properties of the propagation channel. The additive white Gaussian noise (AWGN) Shannon capacity, obtained for thermal receiver noise in the antenna elements, (normalized to transmission bandwidth) expressed in bits per second (bit/s/Hz) [17, 18] is given by

$$C = \log_2(1 + \rho) \tag{3.18}$$

Formula (3.18) assumes that propagation coefficient h is equal to one.

For random fading SISO channel, the capacity is a random and characterized by the average value

$$\overline{C} = \log_2\left(1 + \rho \cdot \overline{|h|^2}\right) \tag{3.19}$$

Figure 3.8a shows the average capacity of the Rayleigh fading channel (RFC) and AWGN Shannon capacity for SISO channel. Propagation Rayleigh channel is presented by two orthogonal independent random Gaussian components with zero mean and unit variance. The amplitude has a Rayleigh distribution and the power $|h|^2$, (which is the square of the amplitude), is distributed exponentially. It is seen from the plots that Shannon and Rayleigh capacity values are different, but not significantly.

3.2.3 MIMO System Capacity

Let us suppose that the propagation channel between M transmitting antenna elements and N receiving antennas is characterized by the matrix with the elements $h_{nm} = \frac{y_n}{x_m}$, which determine the ratio between the signal x_m transmitted by mth antenna element and the signal y_n received by nth antenna (Fig. 3.7a)

$$\mathbf{H} = \begin{pmatrix} h_{11}\ h_{12} \ldots h_{1M} \\ h_{21}\ h_{22} \ldots h_{2M} \\ \ldots\ldots\ldots\ldots\ldots\ldots \\ h_{N1}\ h_{N2} \ldots h_{NM} \end{pmatrix} \tag{3.20}$$

The transmitted voltage signals can be described by vector $\mathbf{X} = \{x_1, x_2 \ldots x_M\}$ and the received voltage signals are given by vector $\mathbf{Y} = \{y_1, y_2 \ldots y_N\}$. In reality, values x_n and y_n are functions of time *t*, but for simplicity and without restriction of generality, index *t* is omitted.

Suppose also that the signal received by any antenna element contains additional additive white Gaussian noise with statistically independent identically distributed components for different elements. White noise can be presented as a vector $\mathbf{N} = \{n_1, n_2 \ldots n_N\}$. The propagation channel is described in the matrix form

$$\mathbf{Y} = \mathbf{H} \cdot \mathbf{X} + \mathbf{N} \tag{3.21}$$

Generally, for equally divided power among the *M* transmit antennas, the capacity value can be estimated using the following formula [18]

$$C = \log_2\left(\det\left(\mathbf{I} + \frac{\rho}{M} \cdot \mathbf{H} \cdot \mathbf{H}^*\right)\right) \tag{3.22}$$

where **I** is identity matrix of dimension *N*, and ρ/Mcharacterizes signal-to-noise ratio at the receive element. It should be noted that Eq. (3.22) holds for a deterministic propagation channel.

In a case of $M \times M$ uncoupled Shannon channels Eq. (3.22) can be simplified [19].

$$C = M \cdot \log_2\left(1 + \frac{\rho}{M}\right) \tag{3.23}$$

Formula (3.23) shows that the MIMO system, which enables the transmission of multiple data stream simultaneously on the same frequency, can increase the bandwidth efficiency by the number of data streams employed. Figure 3.8b presents graphs calculated according to the formula Eq. (3.23) as a function of signal-to-noise ratio ρ expressed in the dB scale. From this graph it is straightforward to see that the channel capacity increases with the number of antennas at the transmitter and the receiver sides. For example, for high signal-to-noise ratio, the channel capacity of a MIMO (4×4) system is two times higher than the channel capacity of a MIMO (2×2) system, and the channel capacity of a MIMO (2×2) system is also two times higher than the channel capacity of a SISO system.

When the propagation channel is random (stochastic), the capacity is random, too. The mean (ergodic) capacity can be defined in this case [18] using the expectation over the channel matrix in Eq. (3.22)

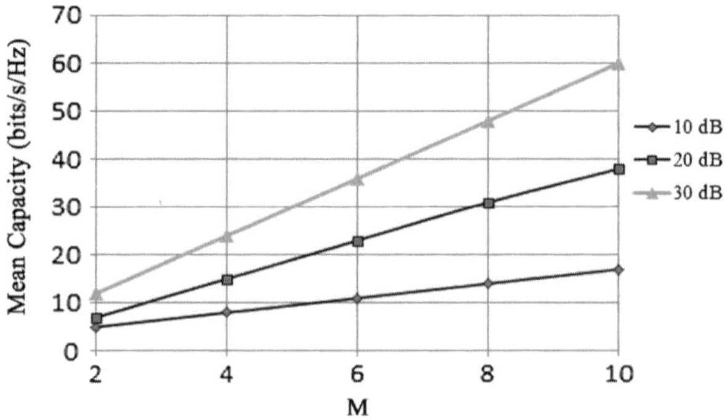

Fig. 3.9 Average capacity values for equal number of transmitting and receiving channels M, i.i.d. MIMO Rayleigh channel

$$\overline{C} = \log_2\left(\det\left(\mathbf{I} + \frac{\rho}{M} \cdot \overline{\mathbf{H} \cdot \mathbf{H}^*}\right)\right) \tag{3.24}$$

In a case of independently, identically distributed (i.i.d.) MIMO Rayleigh fading channels, simplified approximations for capacity $\overline{C}$ held in the paper [20]. Complex channel gain h_{ij} is modeled as an i.i.d. Gaussian random variable with zero mean and unit variance. Figure 3.9 presents the ergodic capacity curves based on the formula approximation calculations. Plots are drawn as a function of the signal-to-noise ratio for different number of channels ($M = N$).

Expression (3.22) can be simplified for the SIMO ($M = 1$) scenario when the branch with the highest SNR is selected as the output signal [17]

$$C_{\mathrm{SIMO}} = \log_2\left(1 + \rho \cdot \max\left(|h_i|^2\right)\right). \tag{3.25}$$

The components $h_i(\mathrm{i} = 1, 2\ldots\mathrm{N})$ determine the propagation channel coefficients between transmit antenna and receive array elements. For combined diversity system capacity is expressed as

$$C_{\mathrm{SIMO}} = \log_2\left(1 + \rho \cdot \sum_{i=1}^{N} |h_i|^2\right) \tag{3.26}$$

For the MISO system without channel knowledge

$$C_{\mathrm{MISO}} = \log_2\left(1 + \frac{\rho}{M} \cdot \sum_{i=1}^{N} |h_i|^2\right) \tag{3.27}$$

3.2.4 *Correlation Effect and MIMO Capacity Value*

MIMO systems in multipath Gaussian uncorrelated fading environment provide a large increase in capacity compared to a single antenna. However, in many practical situations, correlation among the antenna elements in realistic environments due to the poor scattering conditions, results in capacity decrease and, finally, when the correlation coefficient equals to unity, no advantage is obtained by the MIMO architecture. The fading correlation depends on the characteristics of the propagation channel between the transmitting and receiving antenna systems (spatial correlation) and on the mutual array elements coupling effect, which is determined by the scattering (coupling) array matrix **S**-parameters described in Chap. 2. The effect of signal correlation on the MIMO systems has been studied by Monte Carlo simulation in [21] and analytically [22–25].

For correlated Rayleigh channels, a separable correlation model is often assumed [24], and matrix **H** in Eq. (3.24) can be written as

$$\mathbf{H} = \mathbf{F}_R \mathbf{U} \mathbf{F}_T \tag{3.28}$$

where $\mathbf{F}_R$ and $\mathbf{F}_T$ represent the instantaneous correlation matrices from the receiver and transmitter sides, respectively. The matrix **U** is an $N \times M$ complex channel gain matrix containing i.i.d. complex Gaussian values with unit variance. Loyka [25] presents simple explicit capacity expression as a function of the correlation coefficient r for 2×2 MIMO system

$$\overline{C} = \log_2\left(1 + \rho + \left(1 - |r_{12}|^2\right) \cdot \left(\frac{\rho}{2}\right)^2\right) \tag{3.29}$$

The ratio (3.29) is obtained under condition of equal received power by each antenna element.

When $r_{12} = 0$ (uncorrelated channels), Eq. (3.29) reduces to Eq. (3.23) for $M = 2$, when $r_{12} = 1$ (completely correlated channels, Eq. (3.29) is a classical Shannon formula obtained for single propagation channel.

Figure 3.10a shows 2×2 channel capacity Eq. (3.29) versus the absolute value of correlation coefficient $|r_{12}|$for three values of the signal-to-noise ratio ρ: 10 dB, 20 dB, and 30 dB.

Rough analytical estimation of the channel capacity function [23] for the system with M array elements ($M \geq 2$) from the transmitter and the receiver side leads to the plots shown in Fig. 3.10b. Analysis is based on the assumption that the MIMO system operates in Rayleigh-fading environment, with uncorrelated signals at the transmit side, with correlated signals at the receiving antennas, and exponential model for correlation coefficients

$$r_{ij} = r^{-|i-j|} \text{for } i,j = 1, 2 \ldots M \tag{3.30}$$

Figure 3.10b shows the mean value of capacity as a function of antenna element number M ($M = N$), for $\rho = 10$ dB and different values of r. Shown plots indicate that the mean capacity increases almost linearly with the number of antennas; the presence of exponential correlation among the receiving antennas only affects the

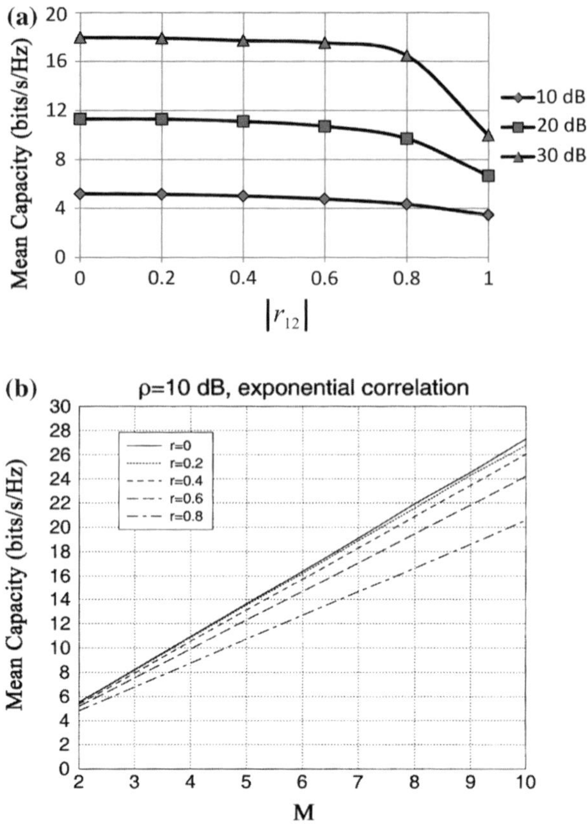

Fig. 3.10 **a** 2 × 2 MIMO system average capacity versus correlation coefficient, Rayleigh fading channel; **b** Average capacity versus number of channels, for exponential model of correlation coefficient [23] © 2003 IEEE

slope of the curves. Furthermore, the reduction of capacity due to the correlation is negligible for values smaller than 0.4.

Real correlation coefficient dependence in a spatial correlation model can be estimated as follows. Let us assume that many independent multipath signals are arriving to the array within $\pm\Delta$ of the mean angle φ of arrival as shown in Fig. 3.11a. If the probability density function of the arrival angles is assumed to be uniform then the signal correlation coefficient between the ith and the kth antenna array elements is

$$r_{in} = \frac{1}{2 \cdot \Delta} \cdot \int\limits_{\varphi-\Delta}^{\varphi+\Delta} \exp[j \cdot k \cdot d \cdot (i-n) \cdot \sin \varepsilon] \cdot d\varepsilon \tag{3.31}$$

where $k = \frac{2 \cdot \pi}{\lambda}$, $d =$ inter-element space, and 2Δ is the angle spread of the incoming multipath waves. For $\Delta = \pi$ ($\varphi = 0$) Eq. (3.31) is a classical expression [26]

$$r_{in} = J_0[k \cdot (i-n)] \tag{3.32}$$

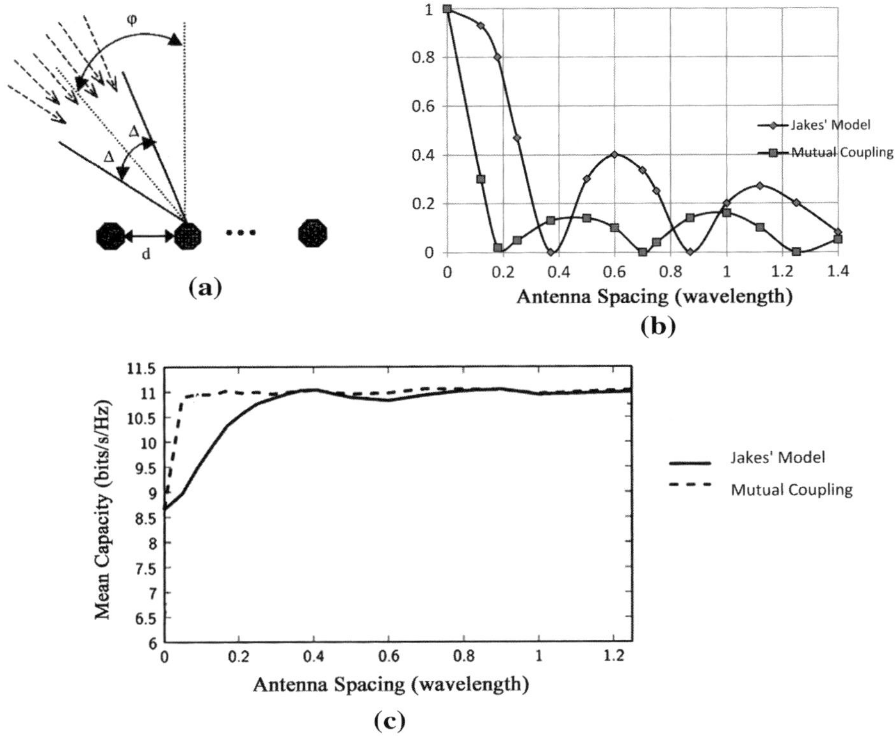

Fig. 3.11 **a** Multipath signals arriving to the linear array; **b** Computer simulation for the envelope correlation coefficient; **c** Mutual coupling effect on the mean capacity for 2 × 2 MIMO system

where J_0 is the zero-order first kind Bessel function.

Figure 3.11b (Jakes model graph) represents computer calculation results of the envelope correlation coefficient as a function of the spacing d between two array elements

$$|r_{12}| = \left|J\left(\frac{2 \cdot \pi \cdot d}{\lambda}\right)\right| \tag{3.33}$$

Figure 3.11c (Jakes model graph, no mutual coupling between elements) plots the average capacity versus spacing for 2 × 2 MIMO system for signal-to-noise ratio equal to 20 dB. The mean capacity was computed over 7,000 realizations for each antenna spacing. It is seen from the graph presented in Fig. 3.11c (Jakes Model) that if the distance between the array elements is more than 0.4 λ, then the mean capacity value is about 11 bits/s/Hz, which is almost the same ($C = 11.34$ bits/s/Hz) as predicted by formula (3.24) obtained for the AWGN MIMO model. When the correlation coefficient is equal to one, the calculation results lead to the classical Shannon formula ($C == 6.6$ bits/s/Hz).

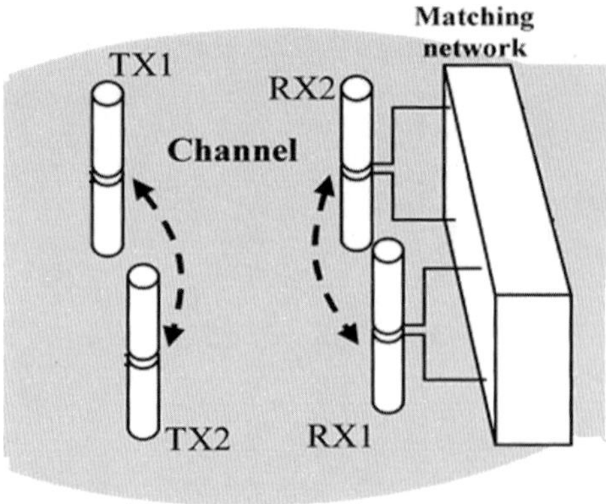

Fig. 3.12 Concept of mutual coupling effect for 2 × 2 MIMO system

3.2.5 Mutual Coupling and MIMO Capacity

For evaluation of the MIMO system parameters, both antennas and the propagation channel have to be treated simultaneously. Formula (3.33) ignores the mutual coupling between antenna array elements, which means that the adjacent elements with spacing d are considered as a no coupling system. Some papers report higher capacity values for closely spaced elements and the others show opposite dependence [27, 28]. It seems that this conflict is caused by ignoring the fact that the total system capacity depends on a number of array element parameters: mutual coupling, element radiation pattern of the isolated element, efficiency, and matching with feeding line. Higher or lower capacity values (in comparison with non-coupling elements) are, in principle, possible and can be specific for different antenna element types. From this point of view, the complete RF system model for analysis of MIMO arrays is of great interest [29, 30]. Such investigation involves a few RF parts: the transmitting antennas, RF channel, and the receiving antennas. For example, when including the mutual coupling between two parallel half wave dipoles [31], the spatial correlation coefficient is different from the well-known expression (3.33). Figure 3.11b (mutual coupling graph) represents the computer simulation results of the correlation coefficient as a function of the spacing between two half wave dipoles shown in Fig. 3.12.

Jakes model corresponds to the omnidirectional antennas with no mutual coupling elements, and the mutual coupling model corresponds to the measurement of the received signal by one antenna with the other terminated by matched impedance. Figure 3.11c (mutual coupling plot) shows the mean capacity value plot versus the receive antenna spacing [31] obtained with mutual coupling in the Rayleigh fading environment. The mutual coupling coefficients were calculated for two half wave dipoles shown in Fig. 3.12 using the Moment Method. The MIMO system is equipped with similar antenna elements at the transmitter and receiver ends. The transmitting dipoles are spaced at one wave distance. Figure 3.11c takes into consideration the array topology and the propagation channel effect. Mathematical simulation clearly shows that for a small spacing due to the mutual coupling effect mean, the capacity value becomes higher. Similar results are shown in the reference papers [29, 32, 33]. Coupling between elements leads to the following channel matrix $\mathbf{H}_c$ [34]

$$\mathbf{Y} = \mathbf{C}^r \cdot \mathbf{H} \cdot \mathbf{C}^t \cdot \mathbf{X} + \mathbf{N} = \mathbf{H}_c + \mathbf{N} \tag{3.34}$$

where matrix $\mathbf{H}$ is obtained for no coupling elements

The capacity in this case becomes

$$C_{\text{coupling}} = \log_2\left(\det\left(\mathbf{I} + \frac{\rho}{M} \cdot \mathbf{H}_c \cdot \mathbf{H}_c^*\right)\right) \tag{3.35}$$

where the coupling matrices $\mathbf{C}^r$ and $\mathbf{C}^t$ determine coupling coefficients from the receiver and transmitter ends, respectively. Mutual coupling effect can also be characterized by a correlation coefficient [35]. For two element array it can be estimated as

$$\left|r_{12}^{\text{coupling}}\right| = \frac{\left|S_{11}^* \cdot S_{12} + S_{21}^* \cdot S_{22}\right|^2}{\left(1 - \left(|S_{11}|^2 + |S_{21}|^2\right)\right) \cdot \left(1 - \left(|S_{22}|^2 + |S_{12}|^2\right)\right)} \tag{3.36}$$

3.2.6 SIMO Diversity Techniques

3.2.6.1 Main Combining Schemes

Antenna diversity techniques are widely utilized in SIMO topology. Four main diversity combining schemes support three different antenna diversity methods: spatial diversity, pattern diversity, and polarization diversity [36, 37]. Spatial diversity employs antennas that are physically separated. Polarization diversity combines antennas with orthogonal or near orthogonal polarizations. Typically reflected signals in the multipath propagation channel undergo polarization changes. By pairing two polarizations, signal processor circuit can maximize SNR of the incoming signal. A pattern diversity technique uses two or more antennas with different radiation patterns. Combining methods can be characterized by the following main features:

(a) Switched Combining

The simplest combining method requires only one receiver for N branches as shown in Fig. 3.13a and one transmitting antenna. The receiver is switched according to the certain (predetermined) algorithm from one to the other branch only when SNR on the current branch is dropped below the predefined threshold.

The threshold can be fixed or variable value depending on the switchable logic. This method is widely used in the automotive industry for FM radio reception. Typically antenna array elements are implemented as printed on car glass array antenna elements.

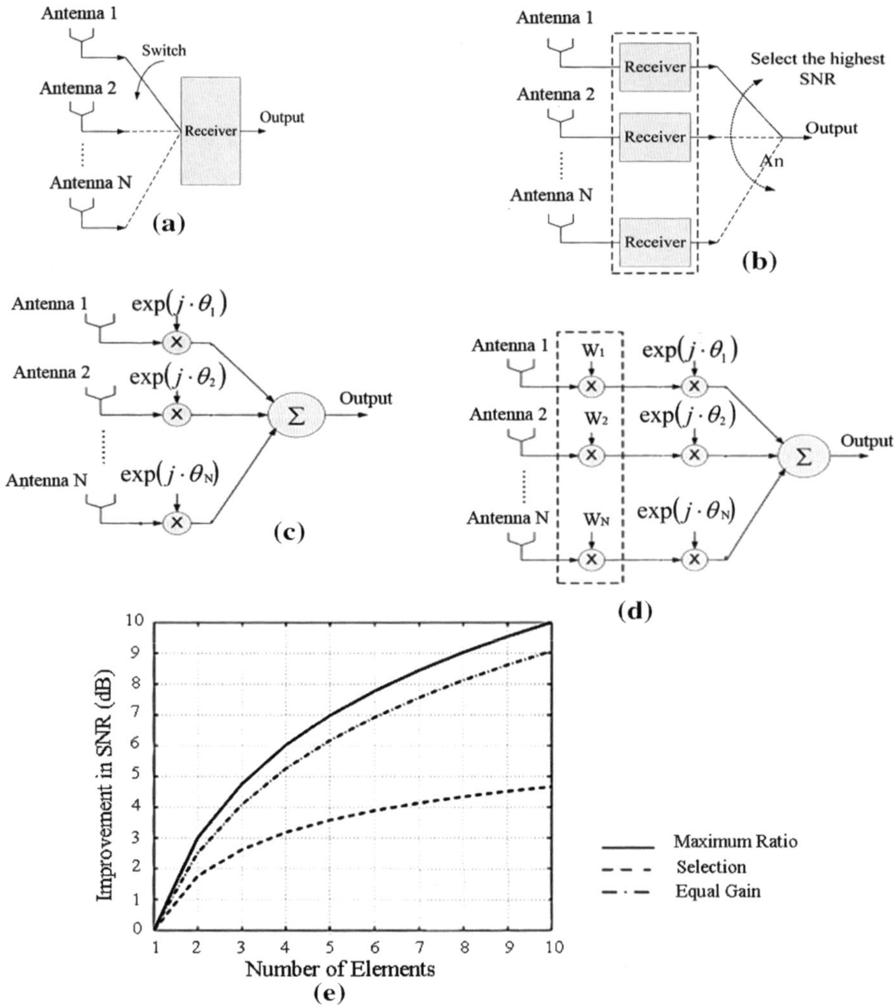

Fig. 3.13 **a** Simplest scanning diversity; **b** Selection combining diversity; **c** Equal gain combining diversity; **d** Maximum ratio combining; **e** Improvement in average SNR for different diversity methods

(b) Selection Combining

The selection combining method presented in Fig. 3.13b is similar to switched combining except that N receivers are required to monitor instantaneous SNR at all branches. The branch with highest SNR is selected as the output signal.

(c) Equal Gain Combining

In order to improve SNR at the output, the signals from all branches are combined to form the output signal as shown in Fig. 3.13c. The signals from different branches are not in phase. Therefore, signal that passes through each branch is multiplied by a complex weight with phase ϑ_i ($i =$ branch number), and all output signals have the same phase combining coherently.

(d) Maximum Ratio Combining

In order to maximize the SNR at the output, each ith signal is applied with a weight w_i before all signals are combined coherently (Fig. 3.13d). A weight factor is proportional to the signal amplitude. That is, branches with a strong signal are further amplified, while weak signals are attenuated.

Comparing all combining methods we can conclude that:

(1) Switched combining technique is the simplest, requiring only one receiver and power measurements;

(2) Selection combining requires only power measurements, however, it uses multiple receivers, the gain in signal-to-noise ratio in comparison with one receiver topology is about ln (N), where $N =$ number of the receive branches;

(3) Maximum ratio combining is optimal in SNR sense, the increase of the SNR in comparison with SISO method is equal to N, however, requires knowledge of the phase and amplitude channel parameters;

(4) Equal gain combining is not optimized as a maximum gain combining method. However, it only requires information about phase channel parameters and, therefore, may be a good tradeoff for the implementation.

Estimation plots of Fig. 3.13e summarize improvement in average SNR versus the number of receiving branches for a few combining techniques.

3.2.6.2 Diversity Gain

Diversity gain is one of the main parameters that determine the efficiency of the diversity system. Diversity gain quantifies the improvement in the signal-to-noise ratio of a received signal that is obtained using different receiver branches. Diversity gain for a given cumulative probability p is given by

$$G_{\mathrm{div}} = \gamma_{div}(p) - \gamma_1(p) \tag{3.37}$$

where γ_{div} is the SNR with diversity and γ_1 is the signal-to-noise ratio of the single branch without diversity combining. Thus, diversity gain is the improvement in the signal-to-noise ratio due to the diversity technique, for a given cumulative probability or reliability.

Diversity gain can be estimated using cumulative distribution function (cdf) for Rayleigh signal which for one channel is given by

$$P_r(\mathrm{SNR} < \mathrm{SNR}_0) = 1 - e^{-\frac{\mathrm{SNR}_0}{\mathrm{SNR}_{\mathrm{av}}}} \tag{3.38}$$

where SNR is the instantaneous signal-to-noise ratio, SNR_{av} is the mean SNR, and $P_r(\mathrm{SNR} < \mathrm{SNR}_0)$is the probability that the SNR will fall below the given threshold, SNR_0.

For a selection combiner with N independent branches, assuming that the N branches have independent signals and equal mean $\mathrm{SNR}_{\mathrm{av}}$, and the probability of

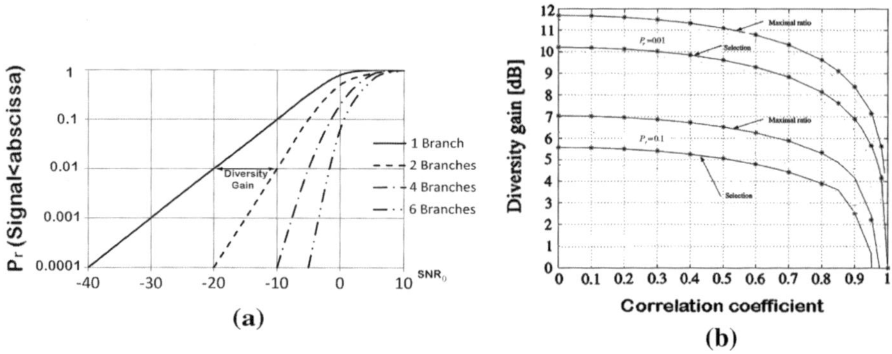

Fig. 3.14 **a** Cumulative distribution function as a function of SNR for different branch numbers, **b** Diversity gain versus correlation coefficient for maximum ratio combining and selecting diversity, two receiving antennas [38] ©IEEE 2004

all branches having a SNR below SNR_0 is equivalent to the probability for a single branch raised to the power N as:

$$P_r(\mathrm{SNR} < \mathrm{SNR}_0) = \left(1 - e^{-\frac{\mathrm{SNR}_0}{\mathrm{SNR}_{\mathrm{av}}}}\right)^N \tag{3.39}$$

Figure 3.14a demonstrates function $P_r$calculated by formula (3.39) for different number of diversity branches. Value SNR_0 is expressed in dB scale. Using these curves, it can estimate the diversity gain for the certain outage probability.

Figure 3.14b [38] shows diversity gain curves as a function of the correlation coefficient$|r_{12}|$ for two different probability values $P_r = 0.1$ and 0.01 for two different combing methods: selection combining and maximum ratio combining methods. These plots, which are obtained for two receive branches, allow estimating diversity gain according to the known value of the correlation coefficient between them.

3.3 High Resolution Processing Algorithms for Direction Finding (DF) Applications

Beamformer techniques described in Chap. 2, which based on the electronically or mechanically scanning antenna arrays, are considered as a fundamental to direction finding (DF) applications. However, the accuracy of these methods drops dramatically when two incident waves are within the main beam of the antenna array. This problem can be overcome by utilizing high resolution DF techniques, for example, Bartlett, Capon, Multiple Signal Classification (MUSIC), or Estimation of Signal Parameters via Rotational Invariance Techniques (ESPIRIT), which are based on the estimation of cross correlation of the signals received by

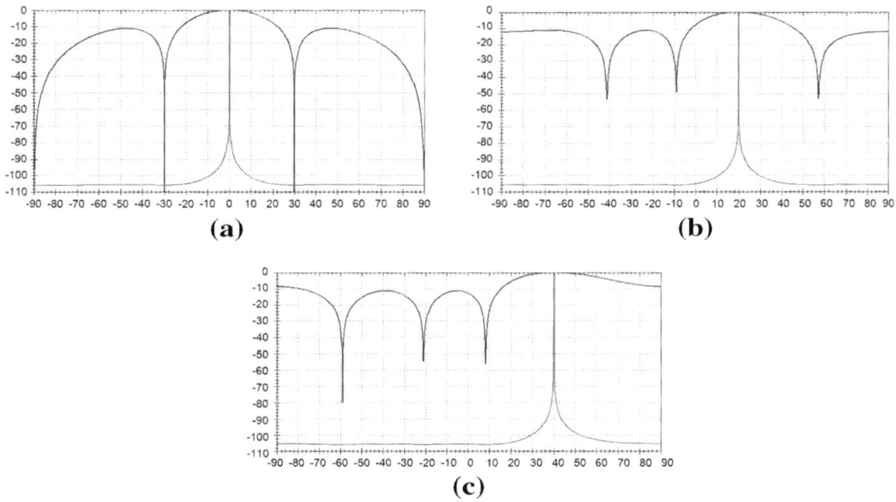

Fig. 3.15 Simulation results for MUSIC algorithm, four-element linear array with 0.5λ spacing, SNR = 20dB, one source with the following angle direction: **a** 0°; **b** 20°; **c** 40°

different array elements [39, 40]. Among all techniques, the most exploited is the MUSIC method. Detailed mathematical explanation of this method is beyond the scope of this book. However, we present short review of the high resolution MUSIC technique.

Let us consider an N element linear array with equally spaced isotropic antenna elements. Assume that each element has its own receiver that is able to record the complex envelope of the incoming signal, and the number of waves impinging on the array is Q. Data recorded by nth element at t_s time moment $(s = 1, 2, 3 \ldots S)$is given by

$$x_n(t_s) = \sum_{q=1}^{Q} A_q(t_s) \cdot e^{j \cdot k \cdot d \cdot n \cdot \sin \theta_q + j \cdot \varepsilon_q(t_s)} + n_n(t_s) \tag{3.40}$$

where $A_q(t_s)$ = time variable amplitude of the incident wave number q applied to the array, $\varepsilon_q(t_s)$ = time variable phase of the source wave, and $n_n(t_s)$ = receiver noise of the nth element with average equal to 0 and variance N. Cross correlation matrix element R_{mn} between the received signals x_m and x_n can be written as

$$R_{mn} = E\left[x_m(t_s) \cdot x_n^*(t_s)\right] \tag{3.41}$$

Temporal averaging $E[\cdot]$ over S snapshots (samples) in Eq. (3.41) is determined by

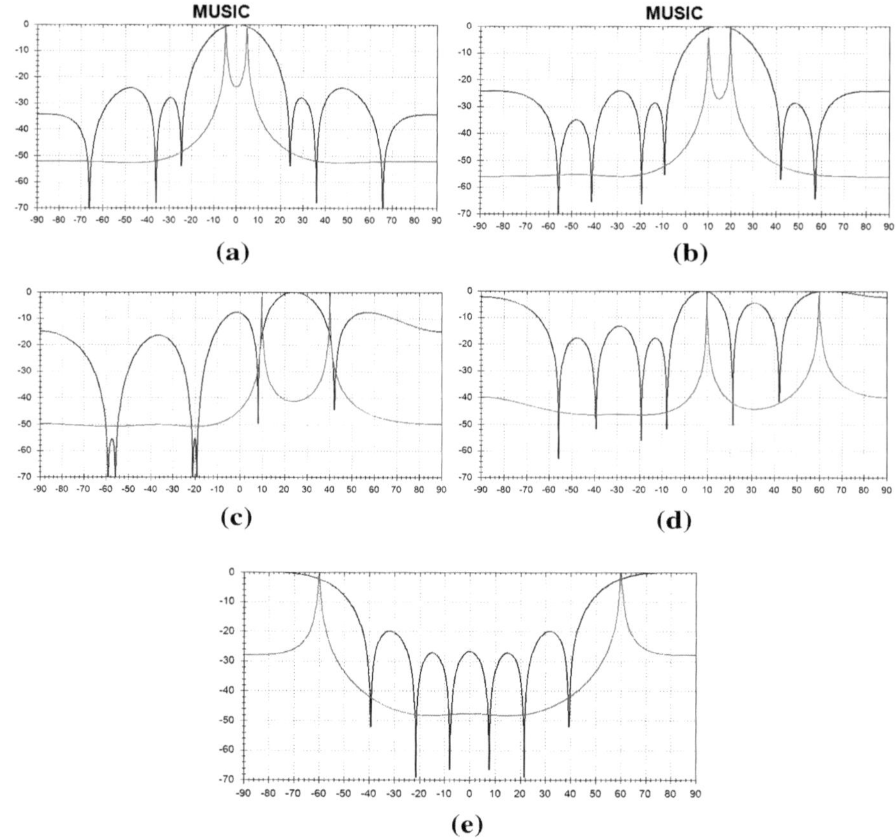

Fig. 3.16 Simulation results for MUSIC algorithm, four-element linear array with 0.5λ spacing, SNR = 20dB, two sources with the following angle direction: **a** 0°and 7°; **b** 10° and 20°; **c** 10° and 40°; **d** 10° and 60°; **e** −60° and 60°

$$R_{mn} = \frac{1}{S} \cdot \sum_{s=1}^{S} x_m(t_s) \cdot x_n^*(t_s) \tag{3.42}$$

Cross correlation matrix **R** consisting of the elements (3.42) can be eigenvalue decomposed as shown in reference paper [39]. The matrix **R** belong to orthogonal signal subspace and noise subspace is given by

$$\mathbf{R} = \widehat{\mathbf{V}}_Q \cdot \Lambda_Q \cdot \widehat{\mathbf{V}}_Q^* + \widehat{\mathbf{V}}_{\text{noise}} \cdot \Lambda_{\text{noise}} \cdot \widehat{\mathbf{V}}_{\text{noise}}^* \tag{3.43}$$

where $\Lambda_Q = \{\lambda_1, \lambda_2 \ldots \lambda_Q\}$ and $\widehat{\mathbf{V}}_Q = [\mathbf{V_1}, \mathbf{V_2} \ldots \mathbf{V}_Q]$ determine eigenvalues and corresponding eigenvectors from signal subspace, while $\Lambda_{\text{noise}} = \{\lambda_{Q+1}, \lambda_{Q+2} \ldots \lambda_N\}$ and $\widehat{\mathbf{V}}_{\text{noise}} = [\mathbf{V}_{Q+1}, \mathbf{V}_{Q+2} \ldots \mathbf{V}_N]$ are related to the noise space. It is

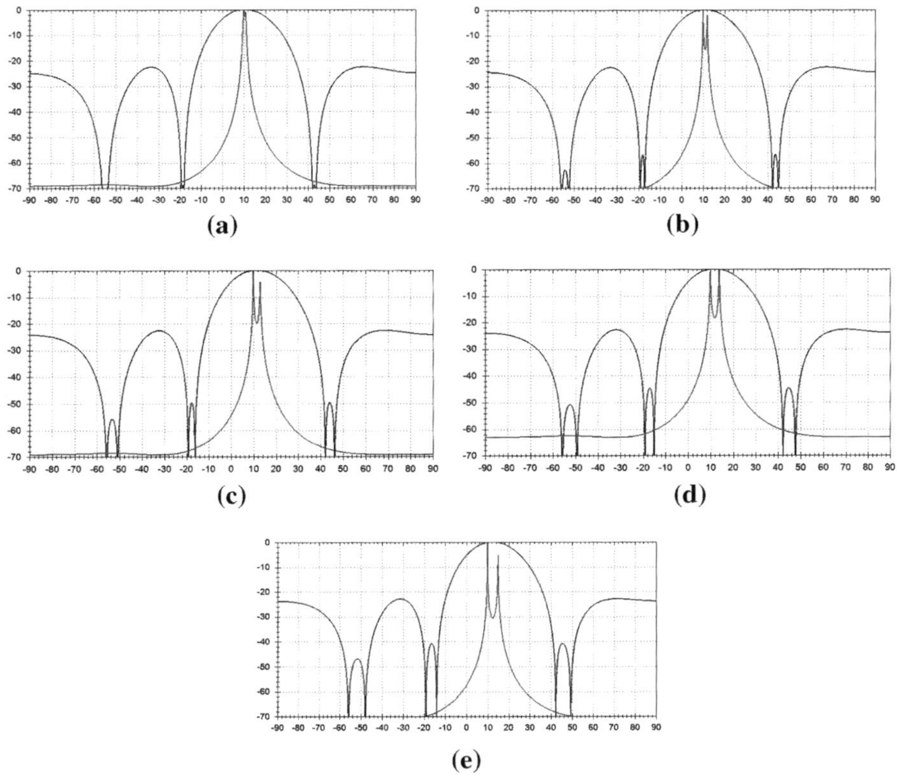

Fig. 3.17 Simulation results, estimation of angle resolution for MUSIC technique, two sources, SNR = 20dB, **a** 10° and 11°; **b** 10° and 12°; **c**10° and 13°; **d**10° and 14°; **e** 10° and 15°

shown [39] that the dimension of the signal subspace is Q, while the dimension of the noise subspace is N–Q. Eigenvalues from noise subspace are the smallest eigenvalues and equal to N. As it follows from the MUSIC method, the number of signals impinging on the array can be estimated as

$$\widehat{Q} = N - \widehat{k} \tag{3.44}$$

where $\widehat{k}$ = the number of smallest eigenvalues.

Based on the presentation of cross correlation matrix **R**, peaks corresponding to the angle of incoming signals can be found from the following algorithm

$$P_{\mathrm{MUSIC}} = \frac{1}{\mathbf{W}(\theta) \cdot \widehat{\mathbf{V}}_{\mathbf{noise}} \cdot \widehat{\mathbf{V}}^{*}_{\mathbf{noise}} \cdot \mathbf{W}^{*}(\theta)} \tag{3.45}$$

where steering vector $\mathbf{W}(\theta) = \left\{1,\ e^{j \cdot k \cdot d \cdot \sin\theta}\ e^{2 \cdot j \cdot k \cdot d \cdot \sin\theta}\ \cdots e^{(N-1) \cdot j \cdot k \cdot \sin\theta}\right\}$.

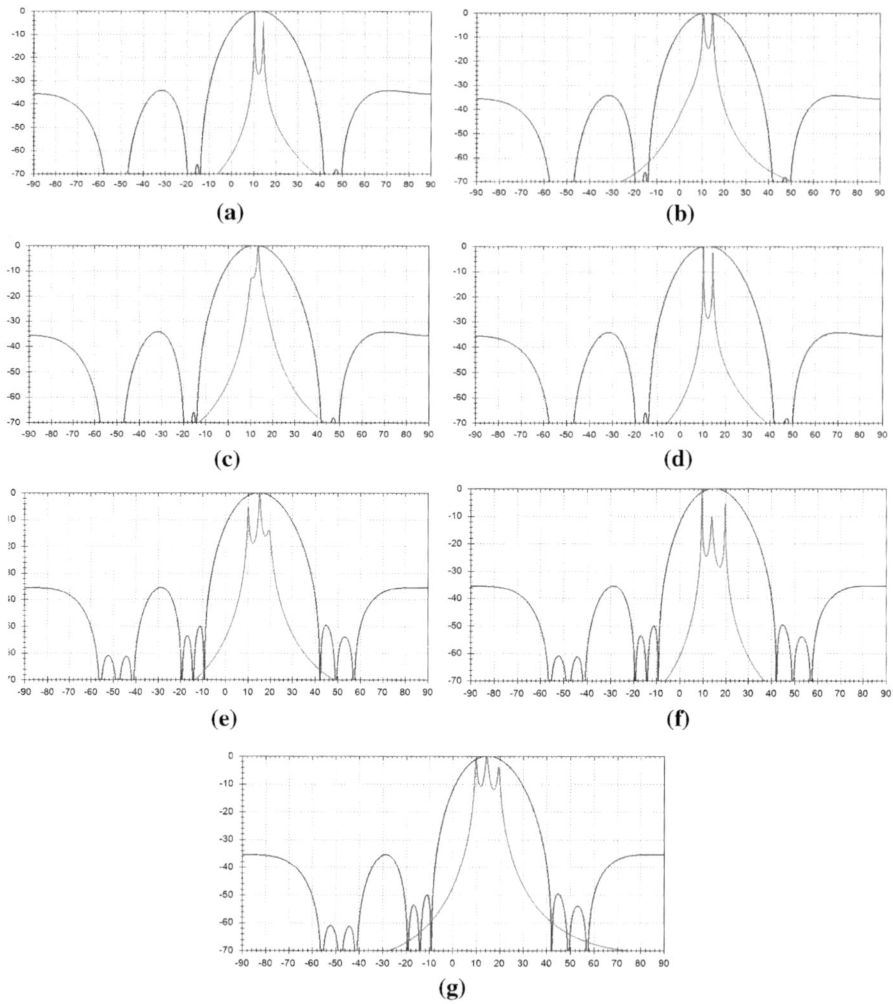

Fig. 3.18 Simulation results, estimation of angle resolution for MUSIC technique, three sources, SNR = 20 dB, source angles for plots **a**, **b**, **c**, **d** are: 10°, 12°, and 15°; source angles for plots **e**, **f**, and **g** are: 10°, 15°, and 20°

Figures 3.15, 3.16, 3.17, 3.18, 3.19 present simulated radiation patterns of four isotropic element linear antenna array with the interelement spacing of 0.5λ when using MUSIC algorithm. For comparison all figures show results obtained using traditional beamforming spatial spectrum plots. Calculations are performed for the antenna array illuminated by one, two, and three RF sources with variable angle distances between them. Correlation matrix elements (3.42) are averaged over 10,000 random realizations, and RF waves incoming from different angle directions are noncoherent.

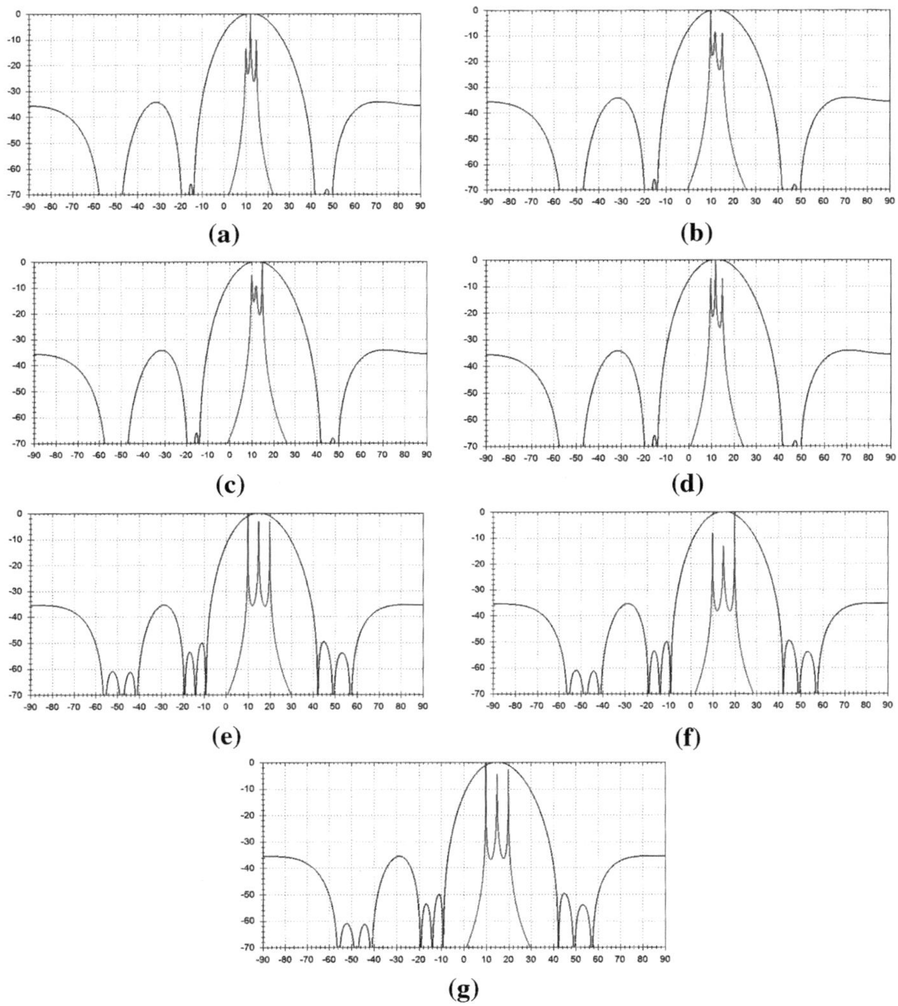

Fig. 3.19 Simulation results, estimation of angle resolution for MUSIC technique, three sources and SNR = 40 dB, source angles for plots **a**, **b**, **c**, **d** are: 10°, 12°, and 15°; source angles for plots **e**, **f**, and **g** are: 10°, 15°, and 20°

Results demonstrated in Figure 3.15a–c are obtained for one incident plane. Figures 3.16 and 3.17 present simulation results for two sources with variable angle distance between them.

As we can see MUSIC algorithm provides much higher resolution than achieved by Rayleigh spatial criterion estimated by the following formula

$$\delta\theta \approx \frac{\lambda}{L} \tag{3.46}$$

where L = antenna linear dimension. According to this criterion, $\delta\theta$ for four-element array with half wave spacing should be 20°.

As it is seen from Fig. 3.17, system with the MUSIC algorithm starts to resolve two sources when the minimum angle between them is equal to 2°, which is ten times less than according to Eq. (3.46) criteria. Finally, Figs. 3.18 and 3.19 illustrate simulation results for three sources for signal-to-noise ratio values: 20 dB (Fig. 3.18) and 40 dB (Fig. 3.19). Plots (a), (b), (c), (d) correspond the same angle positions 10°, 12°, and 15° but with different series of random realizations. Plots named as (e), (f), and (g) are obtained for angles 10°, 15°, and 20°. It is seen that the system with 20 dB signal-to-noise ratio resolves three sources only when angle separation between them is about 5°. However, for signal-to-noise ratio 40 dB, angle resolution is achieved for 2° angle separation between the plane waves.

References

1. Van Atta L (1959) "Electromagnetic Reflection". US patent 2908002, 1959
2. Applebaum S (1976) Adaptive arrays. IEEE Trans Antennas Propag 24(5):585–598
3. Gabriel W (1976) Adaptive arrays—an introduction. Proc IEEE 64(2):239–272
4. Widrow B, Mantey P (1967) Adaptive Antenna Systems. Proc IEEE 55:2143–2159
5. Monzingo R, Haupt R, Miller T (2011) Introduction to adaptive arrays, 2nd edn. Wiley, New York
6. Morgan D (1978) Partially adaptive array techniques. IEEE Trans Antennas Propag 26(6):823–833
7. Kashin VA, Shumilov VF (2006) An adaptive antenna system with digital feedback and analog implementation of weighting coefficients. J Commun Technol Electron 51(8):864–873
8. Steyskal H et al (1986) Methods for null control and their effects on the radiation pattern. IEEE Trans Antennas Propag 34(3):404–409
9. Steyskal H (1983) Simple method for pattern nulling by phase perturbation. IEEE Trans. Antennas Propag 31:163–166
10. Baird CA, Rassweiler GG (1976) Adaptive sidelobe nulling using digitally controlled phase-shifters. IEEE Trans Antennas Propag AP-24:638–649
11. Haupt R (1997) Phase only adaptive nulling with a genetic algorithm. IEEE Trans Antennas Propag 45(6):1009–1015
12. Haupt R (2006) Adaptive antenna arrays using a genetic algorithm. Adaptive and learning systems, IEEE Mountain Workshop, pp. 249–254
13. Vucetic B, Yuan J (2003) Space-time coding. Wiley, Chichester
14. Flam R, Bull J (1988) A circular antenna utilizing phase only adaptive nulling. Antennas and propagation society international symposium. IEEE 3:972–975
15. Kasami H et al (2002) Periodical intermittent interference suppression algorithm for 2.4 GHz band adaptive array. Vehicular technology conference. IEEE 1:435–439
16. Davis RM (1998) Phase-only LMS and perturbation adaptive algorithms. IEEE Trans Aerosp Electron Syst 34:169–178
17. Foschini GJ, Gans MJ (1998) On limits of wireless communications in a fading environment when using multiple antennas. Wirel Pers Commun 6:311–335
18. Telatar I (1999) Capacity of multi-antenna Gaussian channels. Eur Trans Telecommun 10:585–595
19. Jensen M, Wallace J (2004) A review of antennas and propagation for mimo wireless communications. IEEE Trans Antennas Propag 52(11):2810–2824

20. Liang Y (2005) Ergodic and outage capacity of narrowband mimo gaussian channels. Department of Electrical and Computer Engineering, The University of British Columbia, Vancouver, British Columbia, 19 April 2005
21. Winters J (1987) On the capacity of radio communication systems with diversity in a rayleigh fading environment. IEEE J Sel Areas Commun 5(5):871–878
22. Loyka S (2001) Channel capacity of MIMO architecture using the exponential correlation matrix. IEEE Commun Lett 5(9):369–371
23. Shiu D et al (2000) Fading correlation and its effect on the capacity of multielement systems. IEEE Trans Commun 48:502–513
24. Chiani M et al (2003) On the capacity of spatially correlated MIMO rayleigh-fading channels. IEEE Trans Inf Theory 49(10):2363–2371
25. Loyka S (1999) Channel capacity of two-antenna BLAST architecture. Electron Lett 35(17):1421–1422
26. Jakes W (1974) Microwave mobile communications. Wiley, New York, pp 60–65
27. Jungnickel V et al (2003) Capacity of MIMO systems with closely spaced antennas. IEEE Commun Lett 7(8):361–363
28. Volakis J (2007) Antenna engineering book. McGraw Hill, New York
29. Waldschmidt C et al (2004) Complete RF system model for analysis of compact MIMO arrays. IEEE Trans Veh Technol 53(3):579–586
30. Landon D, Furse C (2008) The MIMO transmission equation. Antennas and propagation society international symposium, IEEE
31. Wallace J (2004) Mutual coupling in MIMO wireless systems: a rigorous network theory analysis. IEEE Tran Wirel Commun 3(4):1317–1325
32. Vaughan R (1992) Closely Spaced Terminated Monopoles for Vehicular Diversity Antennas. Antennas and propagation society international symposium, IEEE, vol 2, pp 1093–1096
33. Lau B et al (2005) Capacity analysis for compact MIMO systems. Vehicular technology conference, IEEE, vol 1, pp 165–170
34. Svantesson T, Ranheim A (2001) Mutual coupling effects on the capacity of multielement antenna systems. In: Proceedings of IEEE ICASSP 2001, Salt Lake City, pp 2485–2488
35. Blanch S et al (2003) Exact representation of antenna system diversity performance from input parameter description. Electron Lett 39(9):705–707
36. Fujimoto K, James J (2001) Mobile antenna system handbook, 2nd edn. Artech House, Norwood
37. Dietrich C et al (2001) Spatial, polarization, and pattern diversity for wireless handheld terminals. IEEE Trans Antennas Propag 49(9):1271–1281
38. Mattheijssen R et al (2004) Antenna pattern diversity versus space diversity for use at handhelds. IEEE Trans Veh Technol 53:1035–1042
39. Schmidt R (1986) Multiple emitter location and signal parameter estimation. IEEE Trans Antennas Propag 34(3):276–280
40. Whinnett N, Manikas A (1993) High-resolution array processing methods for joint direction-velocity estimation. Radar and signal processing, IEE Proc F 140(2):114–122

Chapter 4
Simplified Smart Beamforming

4.1 Butler Matrix Topology

Switched beam architecture is considered by many as a cost effective alternative to the fully adaptive array in wireless communication with multiple users. Switched beam systems are technologically simpler and can be implemented using Butler Matrix technology [1–3]. Examples of the Butler matrix for four and eight elements antenna array are shown in Fig. 4.1a, b. The transmission lines are interconnected by a 90° hybrid coupler which schematically is shown in Fig. 4.1c. It provides a 90° phase shift between the signals at the output ports P2 and P3 when input the signal is applied to the port P1 or P4. Figure 4.1d shows the example of the crossover that assures isolation among signals at the crossing lines. It is obtained by cascading two hybrids [4, 5]. Typical achievable insertion losses for the coupler operating at the frequency range 2.4 GHz are about 0.3–0.4 dB, return loss are less than −25 dB, and isolation losses among different ports are less than −20 dB.

As sketched in Fig. 4.1, the number of beam ports equals the number of element ports. The signal injected at any of the beam input terminals excites all radiating elements equally in amplitude, with phase shifts between adjacent antenna elements equal to 180°/N, where N is the total number of the array elements. The radiation pattern of beam number m is given by

$$F(m) = \frac{\sin\left(\frac{N}{2}\cdot\left[k\cdot d-(2\cdot m+1)\cdot\frac{\pi}{N}\right]\right)}{\sin\left(\frac{1}{2}\cdot\left[k\cdot d-(2\cdot m+1)\cdot\frac{\pi}{N}\right]\right)} \tag{4.1}$$

The N beams specified by formula (4.1) cross over at about 4 dB below the beam maximum.

Table 4.1 demonstrates the phase shifts among the four element matrix input and output ports for matrix shown in Fig. 4.1a.

Table 4.2 shows the phase shifts among adjacent antenna ports for different output ports and beam angle directions for the uniform linear array with inter-element spacing equal to $\lambda/2$.

V. Rabinovich and N. Alexandrov, *Antenna Arrays and Automotive Applications*,
DOI: 10.1007/978-1-4614-1074-4_4,

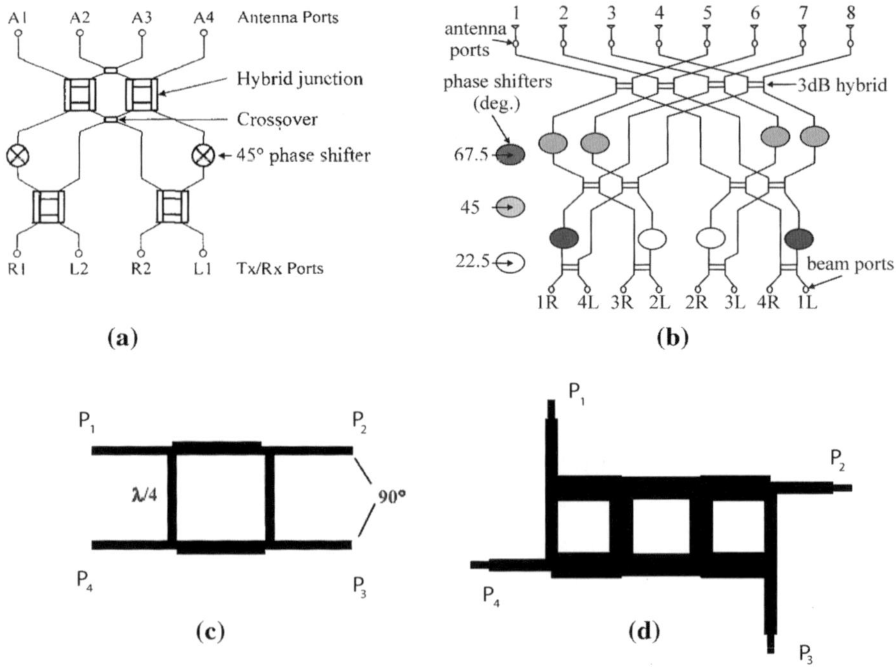

Fig. 4.1 Butler matrix configuration. **a** 4 × 4 topology; **b** 8 × 8 network; **c** Hybrid coupler; **d** Crossover example

Table 4.1 Phase shifts between the input and output ports for 4 × 4 Butler matrix

Ports	A1 (degree)	A2 (degree)	A3 (degree)	A4 (degree)
R1	45	90	135	180
L2	135	0	225	90
R2	90	225	0	135
L1	180	135	90	45

Table 4.2 Angle directions for 4 × 4 Butler matrix

Antenna port	R1 (degree)	L2 (degree)	R2 (degree)	L1 (degree)
Phase shift	45	−135	135	−45
Angle direction	14.48	−48.59	48.59	−14.48

A Butler matrix needs a switching network to direct the appropriate beam to the desired angle location.

Figure 4.2a, which demonstrates eight orthogonal beams produced by the uniform 8 × 8 Butler matrix, shows that central beams have the narrowest beamwidths, whereas the edge beams are the widest. Due to the uniform amplitude distribution across the array elements system has an increased sidelobe level (SLL). It is possible to reduce SLL, by the equal combining of two adjacent ports [6, 7]. It

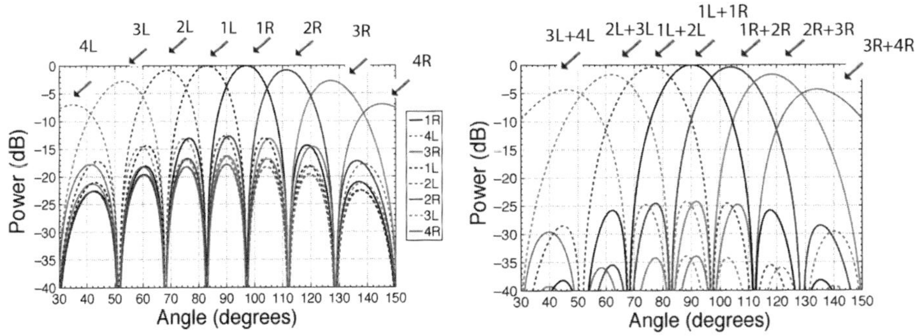

(a) **(b)**

Fig. 4.2 **a** Eight orthogonal beams produced by 8 × 8 Butler matrix. **b** Seven beams with reduced sidelobes for 8 × 8 Butler matrix [7] © 2009 IEEE

is well-known that a cosine or cosine-squared illumination leads to better performance of the sidelobes. From the expression

$$\cos\theta = \frac{e^{j\theta} + e^{-j\theta}}{2} \quad (4.2)$$

it is obvious that cosine-tapered illumination can be regarded as the combination of two uniform illuminations having a progressive phase shifts relative to each other. Consider ports q and r that form excitations $e^{-j\cdot k\cdot d\cdot n\cdot u_q}$ and $e^{-j\cdot k\cdot d\cdot n\cdot u_r}$, where $u_q = \sin\theta_q$ and $u_r = \sin\theta_r$, scan angles are equal to θ_q and θ_r, respectively. Suppose that $u_r = u_q + \delta$. In this case, the combined input

$$e^{j\cdot k\cdot d\cdot n\cdot\left(u-u_q\right)} + e^{j\cdot k\cdot d\cdot n\cdot(u-u_r)} = 2\cdot\cos(k\cdot d\cdot n\cdot\delta/2)e^{j\cdot k\cdot d\cdot n\cdot\left(u-u_q-\delta/2\right)} \quad (4.3)$$

where $u = \sin\theta$, and $\theta =$ observation angle. We see that the combined beam is pointing between two original beams and the amplitudes at the elements have cosine function from the central element to the edges. Calculations show that combining of the beams 3L–4L leads to the maximum SLL equal to −21. 4 dB; 2L–3L → −22.6 dB; 1L–2L → −24.2 dB; 1L–1R → −25.8 dB. However, due to the suppression of the sidelobes, beamwidth increases up to 4° for the combination 1L–1R. Hence, each of the combined patterns has SLL <20 dB, and the maximum of each resultant beam lies approximately at the middle of the two adjacent generating beams shown in Fig. 4.2a. The seven corresponding beams are produced by the combining of two adjacent generating beams, as shown in Fig. 4.2a, are presented in Fig. 4.2b.

4.2 Sectored Antenna Array

The other type of beam switch directional antenna, which can cover 360° with narrow beam, is a sector antenna array (Fig. 4.3) that can utilized for a base station system [8, 9].

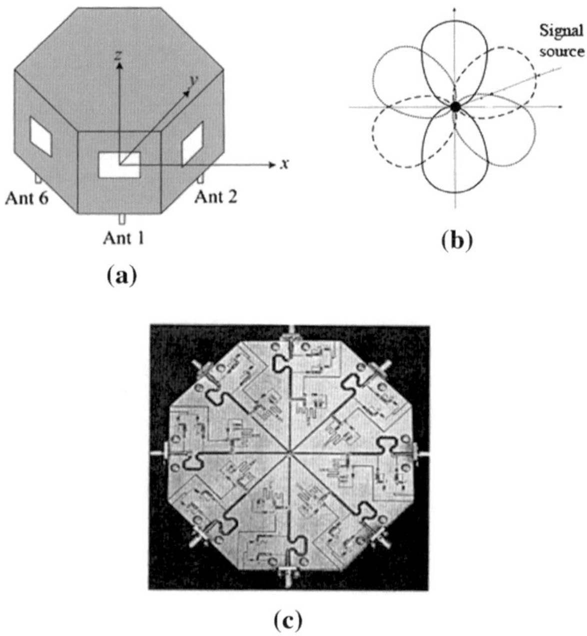

Fig. 4.3 **a** Concept of sectored array configuration. **b** Overlapped radiation patterns. **c** Switching network example [9] ©2001 IEEE

Figure 4.3a demonstrates the concept of the sector configuration considering of six antennas. Each antenna presents an antenna array, producing the directional radiation pattern in the certain angle sector. The patterns of the directional antennas are partially overlapped, as shown in Fig. 4.3b, thus, the system provides 360° coverage. The maximum gains of the adjacent antennas are separated 60° from each other. As shown in Fig. 4.3b, the signal from the incoming source can be received by a few directional antennas but with different signal strength. The simplest algorithm for (direction of arrival) DOA estimation is to choose the strongest signal and identify the DOA as the angle sector with corresponding maximal received power. The system requires a switching network, which provides a successive connection directive antenna with a single receiver. An example of the network with a pin diode switching system built for eight antenna elements covering 360° is shown in Fig. 4.3c.

4.3 Array with Parasitic Elements (ESPAR)

The alternative design to the electronically steerable phased array is the electronically steerable passive antenna radiator (ESPAR) array [10, 11] with electrically controlled parasitic reactive elements. The ESPAR system is a switched beam antenna that steers the radiation pattern within the required angle sector and provides adaptation to the interference sources with low computational cost.

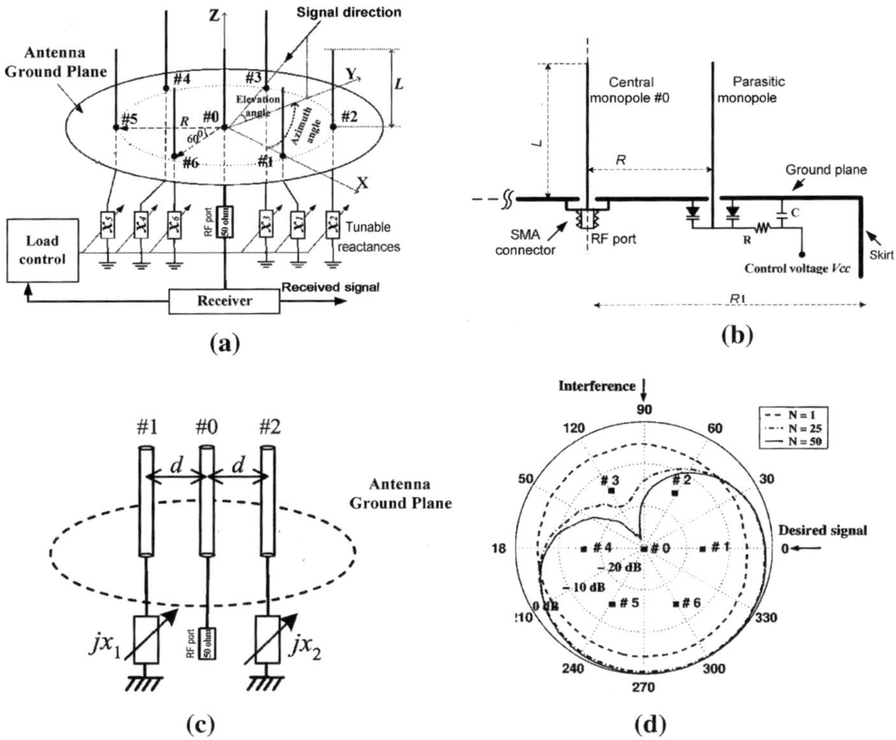

Fig. 4.4 **a** ESPAR array concept sketch. **b** Block diagram of individual reactance circuit. **c** ESPAR linear array configuration. **d** Radiation pattern of ESPAR array with adaptive control

The concept of reactively controlled directive arrays was introduced by Harrington [10], who demonstrated that an antenna array system consisting of one active dipole and a number of passive parasitic elements, that are a short distance from the central element and loaded with the reactive grounded impedances, has a directional radiation pattern. He found that, the beam angle position of the ESPAR array depends on the parasitic elements reactive impedances and can be determined by the matching circuit with electrically controlled varactor diodes. ESPAR antenna systems are expected to play an important role in future wireless communications. They have a couple of advantages over traditional electronically steerable phased arrays: there are no transmission lines to the individual elements, the excitation of elements being accomplished by electromagnetic interaction; elements are relatively closely spaced to provide sufficient electromagnetic interaction, which is a positive factor for antennas mounted on the car roof where the compact factor is very important. Figure 4.4a shows a classic ESPAR circular array configuration example with seven monopole elements (central is an active and six are passive) and Fig. 4.4b demonstrates a typical reactance circuit with one parasitic element. Figure 4.4c demonstrates linear array configuration with one active element and two parasitic monopoles. The length of each monopole and the radius of the circular grid are usually closed to one-quarter wavelength of the RF signal. The vector of the current distribution $\mathbf{W}$ at the ESPAR array, which determines the radiation pattern, is given by [12, 13]

$$\mathbf{W} = 2 \cdot z_s \cdot (\mathbf{Z} + \mathbf{X})^{-1} \cdot \mathbf{u}_0 \tag{4.4}$$

where the $(N + 1) \times (N + 1)$ matrix $\mathbf{Z}$ is the mutual impedance matrix, the diagonal matrix $\mathbf{X} = \mathrm{diag}[z_s, jx_1, \ldots, jx_N]$ is the reactance matrix depending on the desired beam angle position, $\mathbf{u}_0 = [1, 0, 0, \ldots, 0]^T$ is the $(N + 1)$ component vector, $z_S = 50$ ohm is the receiver input impedance. Superscript -1 is the matrix inverse operator. In this case, the circular array factor of the ESPAR system can be expressed as

$$F_{\mathrm{circular}}(\theta) = \mathbf{W}^{\mathbf{T}} \cdot \mathbf{S}(\theta) \tag{4.5}$$

where $\mathbf{S}(\theta)$ is the steering vector, which for circular array is given by

$$\mathbf{S}(\theta) = [S_n(\theta)] = \left[1, e^{j \cdot \frac{\pi}{2} \cdot \cos(\theta - \Psi_1)}, \ldots, e^{j \cdot \frac{\pi}{2} \cdot \cos(\theta - \Psi_N)}\right]; \quad \Psi_m = \frac{2 \cdot \pi}{M} \cdot (m - 1); \\ m = 1, 2, \ldots, N \tag{4.6}$$

A number of adaptive techniques to provide maximum radiation power toward the desired angle and nulls of the radiation pattern toward the interference are employed in ESPAR array systems [14–17]. One of them is a method based on the maximization of normalized cross correlation coefficient r_{yd} [14] between the desired $d(t)$ and the received $y(t)$ output array signal

$$r_{yd} = \frac{\overline{y(t) \cdot d^*(t)}}{\sqrt{\overline{y(t) \cdot y^*(t)}} \cdot \sqrt{\overline{d(t) \cdot d^*(t)}}} \tag{4.7}$$

The total signal $y(t)$ received by the array includes one desired signal incoming from the angle direction θ_0 and a few unwanted signals applied to the array from angle directions $\theta_1 \theta_2, \ldots, \theta_Q$. The objective function J is chosen as a mean square error that has to be minimized for the optimal solution

$$J = 1 - |r_{yd}|^2 \tag{4.8}$$

The output signal of interference ratio is approximately represented by [14]

$$\mathrm{SINR} \approx \frac{|r_{yd}|^2}{1 - |r_{yd}|^2} \tag{4.9}$$

For digital implementation, time t has to be replaced by time samples t_n or simply n. A few procedures are used to provide minimum of function (4.8), or maximum of correlation function: steepest gradient method [15], method based on simultaneous perturbation stochastic approximation with a maximum cross correlation criterion [16], Hamiltonian algorithm [17], power pattern cross correlation method [14], etc. The algorithm, which maximizes (4.7) or minimizes (4.8), [14, 16] can be presented as follows:

$$\mathbf{x}(v+1) = \mathbf{x}(v) - \mu \cdot \nabla J(v) \tag{4.10}$$

where $\nabla J(v) = \left[\frac{\partial J(v)}{\partial x_1} \frac{\partial J(v)}{\partial x_2} \cdots \frac{\partial J(v)}{\partial x_M} \right]^T$.

Reactance value vector $\mathbf{x}(v)$ and gradient $\nabla J(v)$ are obtained at the vth iteration. Each gradient value can be estimated directly by the perturbation method as $\frac{\partial J(v)}{\partial x_n} \cong \frac{J(v+1) - J(v)}{\Delta x_n}$ when the nth reactance x_n is perturbed for Δx_n value. The step size μ and Δx_n can be chosen as positive constants. Figure 4.4d presents the example of the radiation pattern in the H-plane for the array shown in Fig. 4.4b with the following structured parameters: $L = 0.22\lambda$; $R1 = 2R = 0.48\lambda$, and operating frequency is 2.4 GHz. The design is carried out for maximum gain at the azimuth angle 0° and minimum signal toward the interference source (90°). It is seen that the algorithm for this specific scenario shows cancellation of the interference -20 dB in comparison with the maximum value.

However, variable parasitic element reactance complicates the system substantially; therefore, the idea to replace the variable reactance with a simple switch on/off is very attractive. A number of studies [18–20] have followed upon this work with different antenna configurations and investigated the ability to control the principle direction of the radiation pattern. Figure 4.5a [19] shows the circle antenna design consisting of one central active monopole and rings of parasitic monopoles. They can be in open or grounded positions using pin-diode switches, as schematically shown in Fig. 4.5b. By activating (shorting to the ground) the required number of parasitic elements and the pointing of the radiation pattern should be electronically controlled in the azimuth plane. The driven (active) monopole is a quarter wavelength element, and the parasitic elements are arranged in two concentric circles centered at the $\lambda/4$ monopole. The diameters of the inner and outer circles is approximately $2\lambda/3$ and λ, respectively. There are eight parasitic elements in each circle spaced at 45° intervals. The direction of the maximum radiation pattern in the azimuth plane can be rotated by an activated set of parasitic elements shown in Fig. 4.5c.

Figure 4.6a, b [19] demonstrate the azimuth and elevation radiation patterns of the array configured according to the Fig. 4.5a.

4.4 Single Receiver Antenna Array with Digital Beamforming

The standard digital array beamformer processing technique is based on the simultaneous processing of the time sampled signals recorded by all antenna array elements and, hence require that the number of receivers should be equal to the number of receiving antennas. However, for compact, cheap applications, the use

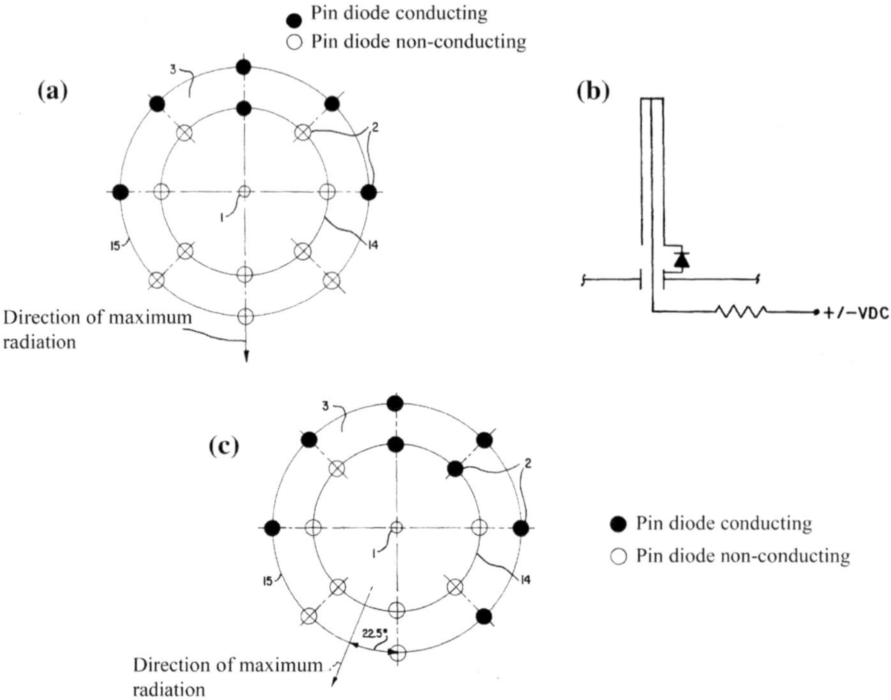

Fig. 4.5 Array with on/off switchable parasitic elements. **a** Array model. **b** Individual element simplified switch circuit. **c** Direction of maximum radiation

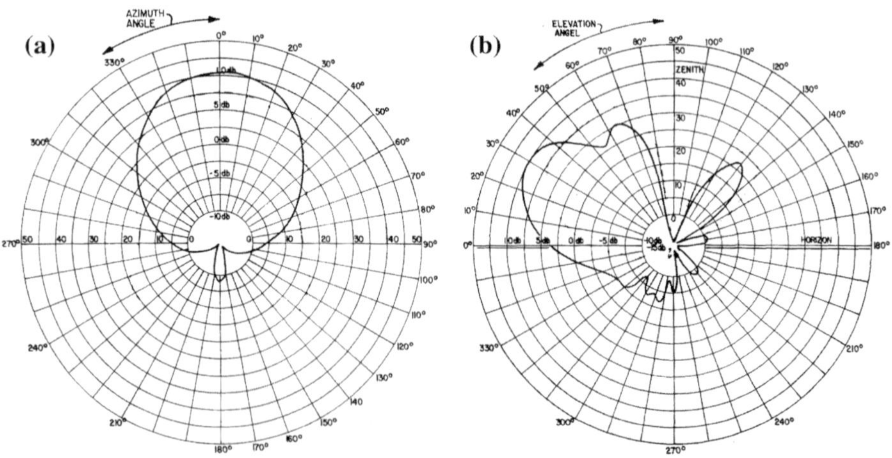

Fig. 4.6 **a** Azimuth radiation pattern for vertical polarization. **b** Elevation radiation pattern

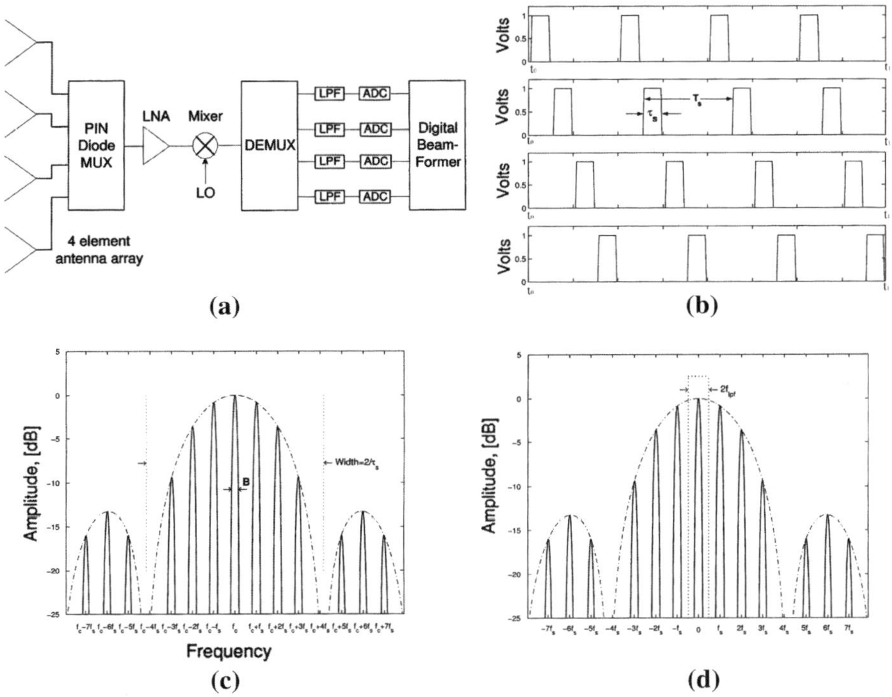

Fig. 4.7 **a** Concept of single receiver array with digital beamforming. **b** Sketch of sequence time diagram. **c** Power spectrum shape. **d** Base-band spectrum

of multiple receivers is impractical. Therefore, single channel techniques are of interest [21–27], particularly in mobile scenarios. One such technique is demonstrated with the sketch shown in Fig. 4.7a.

Method, known as spatial multiplexing of local elements (SMILE) [21], is based on turning on and off individual antenna elements sequentially at a speed above the signal bandwidth. The system presented in Fig. 4.7a includes antenna array elements, a multiplexing circuit with switchers, a single RF channel with a low noise amplifier (LNA), mixer, demultiplexer, low pass filters (LPF), ADC, and a digital signal processor. The multiplexing network is driven by the digital sequence generator (not shown in Figure) so that signals from the array elements are sequentially multiplexed to form a single channel RF output that is connected to the LNA. An example of the digital sequence time diagram is shown in Fig. 4.7b. Figure 4.7c shows the shape of the power spectrum due to the sampling of the signal with carrier frequency f_c and signal bandwidth B. The width of the main spectra lobe is $\frac{2}{\tau_s}$ and the spacing between repeated spectra is $f_s = \frac{1}{N \cdot \tau_s}$, where N is the number of the array elements. To avoid the aliasing effects, the minimum switching rate has to satisfy the following expression (Nyquist sampling theory):

$$f_s \geq B \cdot N \tag{4.11}$$

The analog demultiplexer, driven the same sequence generator, is then used to separate the signals from the different array elements. LPFs are used to reconstruct the continuous signals from the samples. The baseband signal is reconstructed if the cutoff frequency f_{lpf} is selected with the condition

$$\frac{B}{2} < f_{\text{lpf}} < f_s - \frac{B}{2} \tag{4.12}$$

Digitized signals at the output of ADCs are utilized in digital beam processor to calculate antenna array factor. Figure 4.7d demonstrates base band spectrum after LPF.

One problem arises when using the switching network (multiplexing circuit) due to the time delay during the switches among antennas ("time shifting") [25]. The time delay causes additional phase shifts among array elements reconstructed at the ADC outputs. However, these phases are determined by known switching time, and, therefore, can be compensated at the final step when calculating the antenna array factor. Experiments with four element antenna array operating at 5.8 GHz and the sampling rate of 500 k samples per second (or 1.5 times the Nyquist sampling rate) [21] showed a slightly shifted peak of the radiation pattern (around 2° at the angle 20°) due to phase errors caused by the finite switching speed of the pin diode switching network and leakage in the switching pin diodes.

4.5 Time-Modulated Array Configuration

The time-modulated technique synthesizes "time average" tapered amplitude distribution at the uniformly excited array while forming an ultra-low sidelobe radiation pattern [28, 29]. The concept of the time modulated array can be explained with reference to Fig. 4.8a.

If all switches are closed, the array operates as a conventional antenna with uniform amplitude distribution. For the duration of time modulation, each element is connected to the receiver (switched "ON") for a certain period of time proportional to the relative amplitude weights A_n (n = element number).

Another potentially powerful attribute of time-modulated antenna array is the harmonic beam steering. As it is shown in recent publications [30, 31] a sequential switching sequence applied to the elements across the length of a linear array can be used to generate harmonic radiation patterns that are scanned within the certain angle range. Under certain conditions, a designer can combine this technique with low sidelobe pattern synthesis.

The antenna array factor for the linear array consisting of N uniform spacing d and omnidirectional elements is given by:

$$F_{\text{linear}}(\theta, t) = e^{j \cdot 2 \cdot \pi \cdot f_0 \cdot t} \cdot \sum_{n=1}^{n=N} A_n(t) \cdot e^{j \cdot k \cdot d \cdot n \cdot \sin \theta} \tag{4.13}$$

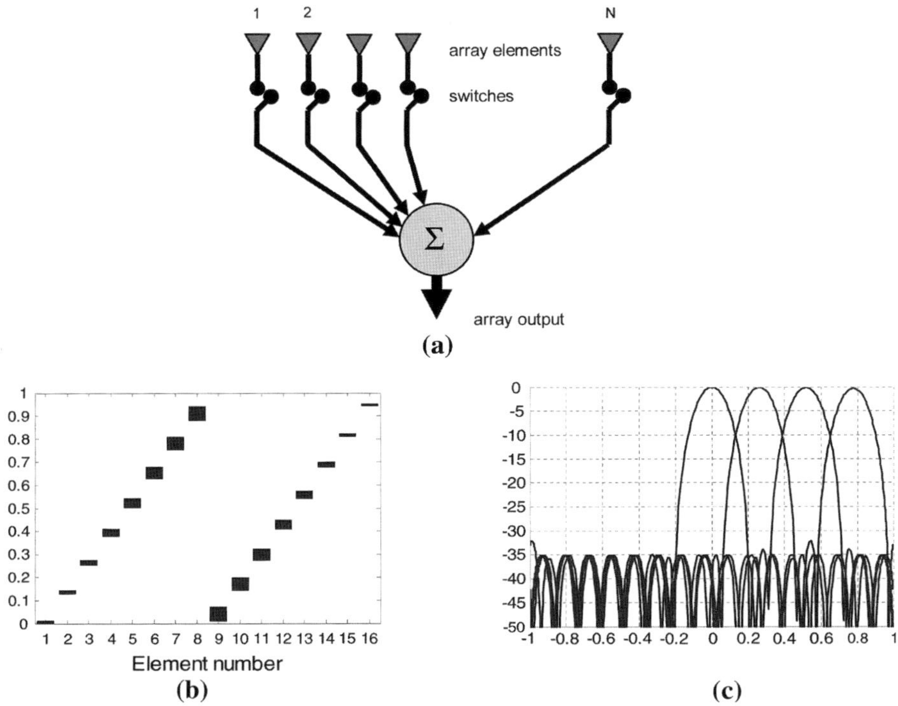

Fig. 4.8 **a** Block diagram of the time modulated array. **b** Element switching sequence required for −35 dB Chebyshev weighting. **c** Simulation results of the multi-beam radiation patterns, [32] ©2010 IEEE

Let us define switched ON/OFF time function

$$A_n(t) = \begin{cases} 1, & \text{if} \quad \tau_{n\text{ON}} \leq t \leq \tau_{n\text{OFF}} \leq T_p \\ 0, & \text{others} \end{cases} \tag{4.14}$$

where $\tau_{n\text{ON}}$ and $\tau_{n\text{OFF}}$ represent the switch-on and switch-off times of element n, respectively, and T_p is the time modulation period (one cycle switching sequence time for all elements). Due to the periodicity of function $A_n(t)$, response (4.13) can be represented through its Fourier series

$$F_{\text{linear}}(\theta, t) = e^{j\cdot 2\cdot\pi\cdot f_0\cdot t} \cdot \sum_{m=\infty}^{m=\infty} \sum_{n=1}^{N} C_m(n) \cdot e^{j\cdot m\cdot\frac{2\cdot\pi}{T_p}\cdot t} \cdot e^{j\cdot k\cdot d\cdot n\cdot\sin\theta} \tag{4.15}$$

$$C_m(n) = \frac{\sin\left(\frac{\pi\cdot m\cdot(\tau_{n\text{OFF}} - \tau_{n\text{ON}})}{T_p}\right)}{\pi\cdot m} \cdot e^{-j\cdot\pi\cdot m\cdot\frac{(\tau_{n\text{ON}} + \tau_{n\text{OFF}})}{T_p}} \tag{4.16}$$

For fundamental harmonic ($m = 0$)

$$F_{\text{linear}}^{m=0}(\theta, t,) = e^{j\cdot 2\cdot\pi\cdot f_0\cdot t} \cdot \sum_{n=1}^{N} \frac{(\tau_{n\text{OFF}} - \tau_{n\text{OFF}})}{T_p} \cdot e^{j\cdot k\cdot d\cdot n\cdot \sin\theta} \tag{4.17}$$

As we see fundamental antenna array harmonic has amplitude weighting coefficients (amplitude distribution along the array elements) proportional to the time pulse durations $\tau_n = \tau_{n\text{OFF}} - \tau_{n\text{ON}}$. If we choose τ_n values corresponding, for example, to the tapered Chebyshev or Taylor amplitude distribution, then fundamental array radiation pattern harmonic will have ultra-low sidelobe levels (less than -30 dB). Fourier coefficients (4.16) may be interpreted as complex weights in which the term

$$w_n = \frac{\sin\left(\frac{\pi\cdot(\tau_{n\text{OFF}} - \tau_{n\text{ON}})}{T_p}\right)}{\pi} \tag{4.18}$$

controls the amplitude of the array elements and the term $\pi \cdot \frac{(\tau_{n\text{ON}} + \tau_{n\text{OFF}})}{T_p}$ introduces a progressive linear phase shift at the harmonic frequencies defined by m. Figure 4.8b [32] shows the element switching sequence required for -35 dB sidelobe Chebyshev weighting function $\tau_n = \tau_{n\text{OFF}} - \tau_{n\text{ON}}$ to scan the first harmonic beam of the 16 element array to 15°. The corresponding array factors at the fundamental frequency and first three harmonic frequencies are shown in Fig. 4.8c, where it is observed that the SLL -35 dB is achieved at both fundamental and first harmonic frequency.

Hence, electronic control of the values $\tau_n = \tau_{n\text{OFF}} - \tau_{n\text{ON}}$ leads to the equivalent weighting of the coefficients (4.18), which provides the concept of the adaptive nulling [33, 34] to the time-modulated antenna arrays. For example, well-known adaptive Applebaum algorithm described in Chap. 3 may be utilized for regulations of amplitude weighting coefficients only. The algorithm is employed dynamically when the interference source angle direction is unknown or weighting amplitude coefficients and subsequent time values τ_n are calculated in advance. The angles of interference sources are predetermined in advance and do not change during the system operation. System operation with an unknown source angle direction requires RF receivers in each array channel with digital control [33], as shown schematically in Fig. 4.9a. This structure can be named as a hybrid analog–digital adaptive beamforming control, because the analog output RF signal is controlled by the digital feedback loop circuit. Figure 4.9b–d shows simulation results of the 8-element isotropic linear array with $\lambda/2$ equal spacing when the interference plane wave comes to the array from the angle equal to 25°. Figure 4.9b corresponds to the initial (before adaptation) radiation pattern, Fig. 4.9c is obtained after adaptation cycle, and Fig. 4.9c demonstrates amplitude weight coefficients. As we can see, when applying amplitude adaptation, weight coefficients are symmetrical relative to the array physical center, and the radiation pattern has two symmetrical dips (25 and $-25°$). A number of papers [35–38] is

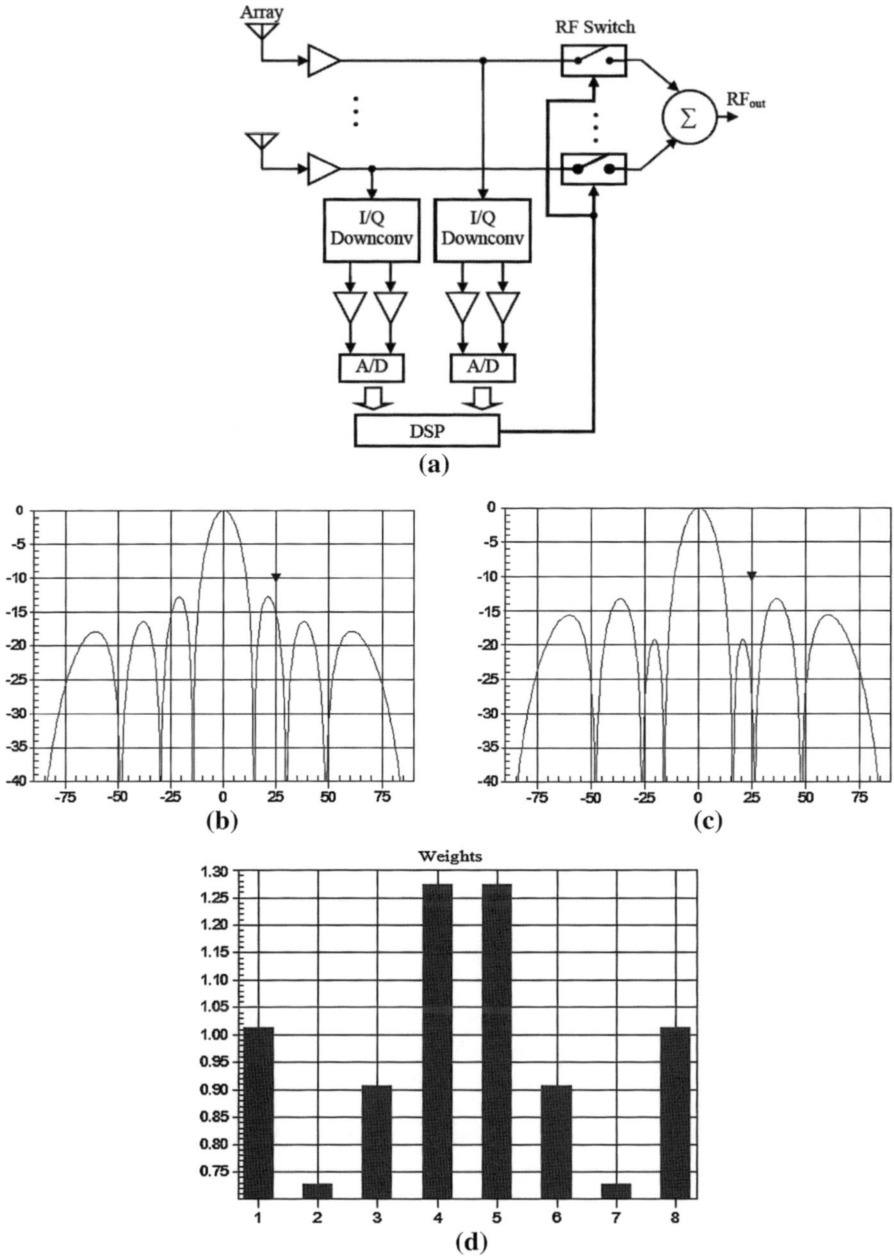

Fig. 4.9 **a** Hybrid analog-digital single receiver adaptive array. **b** Radiation pattern for 8-element linear array with half wave spacing (before adaptation). **c** Radiation pattern after adaptation, interference source angle is 25°. **d** Weight coefficient values

devoted to the amplitude and phase adaptation in the time-modulated arrays. These effective methods based on the signal to noise maximum LMS criteria can be employed for the adaptation of fundamental harmonic as well as for first time-modulated array harmonic.

4.6 Partially Adaptive Small Phased Array with Three-Bit Digital Phase Shifters

Low cost and efficient design is the priority for any commercial system. In this section, we demonstrate simulation results of a small phased array that uses a simple brute force adaptive method to form radiation pattern nulls toward the interference source. In computer science, brute-force search is a trivial but very general problem-solving technique that consists of systematically enumerating all possible candidates for the solution and checking whether each candidate satisfies the problem's statement. Let us apply this technique for an adaptive array with small element numbers and discrete phase shifters. An adaptation criterion is based on minimizing the array output power of the signals incoming from the interference sources. As we noted earlier in Chap. 3, Sect. 3.1.4, Haupt investigated smart genetic algorithm (GA) [39] for the array with digital phase shifters that have minimal phase discrete $\Delta = \pi/32$. In this case, direct brute-force method is impractical. For example, for the array consisting of 20 elements and six bit phase shifters total number of measured power samples for all phase variations is equal to 2^{120}. However, for a vehicular mounted compact array with two or four elements and phase discrete equal to 45°, simple brute-force technique should be reasonable.

Let us assume that we have a linear array with isotropic elements, discrete electronically controllable Q-bit digital phase shifters, and a single receiver. $N + M$ array elements provide beam steering while M antennas additionally provide adaptive phase perturbations. L interference sources with amplitudes I_l ($l = 1, 2, \ldots, L$) arrive at the array from the unknown angle directions $\theta_{l'}$. Q-bit phase shifter has 2^Q different phase states and the phase discrete $\Delta = \frac{2\cdot\pi}{2^Q}$. The measured output power is given by:

$$P = P(\psi_1, \psi_2, \ldots, \psi_M) = \left|\sum_{l=1}^{L} I_l \left(\sum_{n=1}^{N} e^{j\cdot k\cdot d\cdot(n-1)\cdot\sin\theta_l} + \sum_{m=1}^{M} e^{j\cdot k\cdot d\cdot(N+m)\cdot\sin\theta_l + j\cdot\psi_m}\right)\right|^2 \tag{4.19}$$

where each of the phases ψ_m can take any value between 0 and 360° from the following discrete set: Δ, 2Δ ,..., $2\pi-\Delta$. Expression (4.19) is obtained, assuming that the angle of desired source is known, the antenna array consisting of $N + M$ elements electronically is directed to the desired source, and the receiver noise power is much lower in comparison with the output power received by the array (before and after adaptation) from the interference source. Adaptive algorithm is based on minimizing the output power P signal measured by signal strength indicator (RSSI) circuit. It is assumed that the desired signal is not received or transmitted during the adaptation process. During the adaptive process, value P is measured for phase variations from the possible discrete set for each element number m. Hence, the total number of measured power data is equal to $R = 2^{Q \cdot M}$. Among these R measured power values, the computer selects minimal power and corresponding set of M discrete phases $\Psi = \{\psi_1, \psi_2, \ldots, \psi_M\}$. The elements of this set are utilized as the adaptive phase weighting coefficients. For example, for the array with two adaptive elements, and phase discrete 45°, $R = 64$, which is practically reasonable. To prevent signal degradation toward the main beam direction in parallel with phase perturbations, it is estimated radiation pattern toward the desired signal incoming angle θ

$$|F_{\max}|^2 = \left| N + \sum_{m=1}^{M} e^{j \cdot \psi_m} \right|^2 \tag{4.20}$$

If value (4.20) is less than certain level, then corresponding phase set $\Psi = \{\psi_1, \psi_2, \ldots, \psi_M\}$ is replaced by the other one which produces nulling of the interference and reasonable main peak level.

Figure 4.10 shows simulation results of the array consisting of four half-wave spaced isotropic elements, two of them (adjacent elements positioned from one edge) have adaptive phase regulation with a discrete 45°. One interference plane wave arrives at the angle θ. The radiation pattern after adaptation cycle is calculated as:

$$|F(\theta)|^2 = \left| \sum_{n=1}^{N} e^{j \cdot k \cdot d \cdot (n-1) \cdot \sin\theta} + \sum_{m=1}^{M} e^{j \cdot k \cdot d \cdot (N+m) \cdot \sin\theta + j \cdot \psi_m} \right|^2 \tag{4.21}$$

where $N = 2$ and $M = 2$

Figure 4.10a shows a quiescent radiation pattern of four isotropic elements, each of Figs. 4.10b–e represents the radiation pattern after adaptation for the interference source arriving at a different angle, which is shown by the arrow. Adaptive cycle for certain interference angle position consists of measuring 64 output power levels (4.19) for different sets of discrete phases ψ_m ($m = 1, 2$), and choosing phase set $\Psi = \{\psi_1, \psi_2\}$ that produces deep nulling and low main lobe reduction. Table 4.3 demonstrates the final adaptive phase set, the number of iterations after it reached minimum power toward the interference, and corresponding main lobe loss level. Suppression values of sidelobes and main lobe level are presented in comparison with correspondent values of quiescent radiation pattern.

Results demonstrate reasonable quality interference suppression and low loss of the main lobe peak.

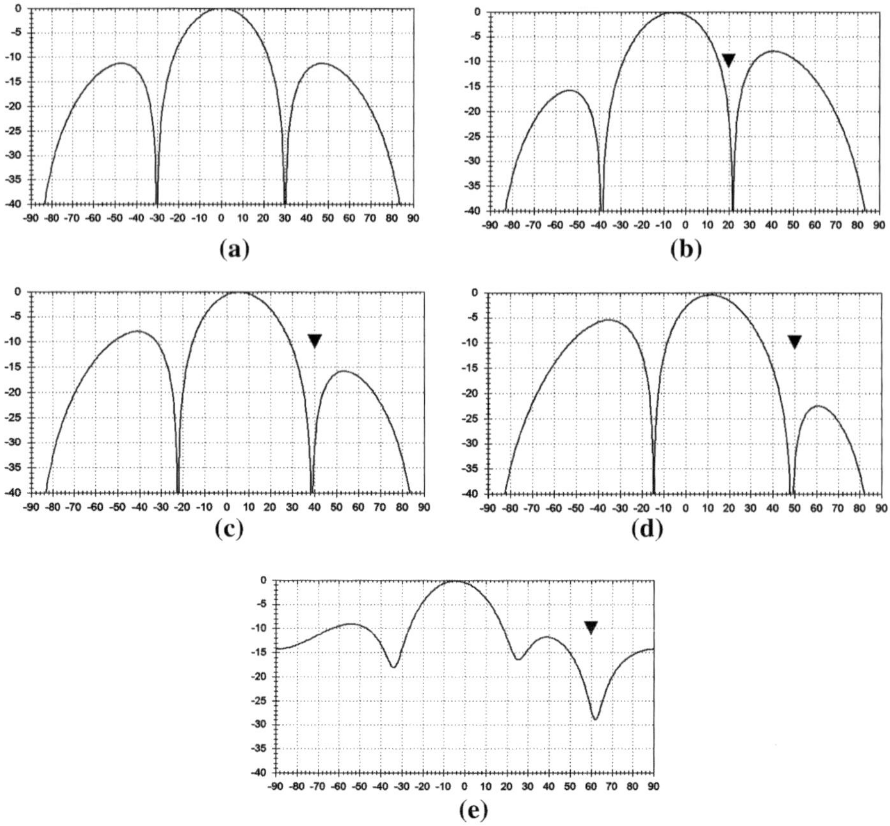

Fig. 4.10 Simulation results of the radiation pattern for the adaptive phase array with three-bit digital phase shifters, total element number 4, two adaptive elements with three-bit phase shifters. **a** Before adaptation. **b** Interference source angle 20°. **c** Interference source angle 40°. **d** Interference source angle 50°. Interference source angle 50°

Table 4.3 Final adaptive phase set, and number of iterations required to obtain minimum power toward the interference source angle

Angle of interference (degree)	Successful step number	Phase for element #1 (degree)	Phase for element #2 (degree)	Sidelobe (dB) level suppression	Main lobe loss level (dB)
20	9	45	45	−13	−0.7
40	63	315	315	−18	−0.7
50	54	270	270	−23	−3
60	1	0	45	−12	−0.5

4.7 A Novel Small Phased Array with Two-Bit Phase Shifters for DOA Applications

Generally, antenna systems for DOA applications use beam steering phased antenna arrays with electronically controlled phase shifters [40] or digital arrays with a processor that provides high-resolution algorithms [41]. Most phase shifters utilized in the electronically steered phased array are digital. It is known [42] that the accuracy of DOA estimation increases for a large antenna array with multi-bit digital phase shifters. Typically, complicated and expensive multi-element arrays are employed only for military applications. DF digital arrays utilize a number of complicated receivers with RF circuits, digital components, and a special calibration system that synchronizes signals to be detected by different antenna elements.

This section describes a direction finding system that includes a low cost, compact, electronically controlled phased antenna array with only two-bit phase shifters. Specific phase variations of phase shifters and the processing algorithm provide improved (in comparison with conventional phased array) resolution of the DOA estimation. Proposed array topology allows estimating amplitude and phasing distribution over the array aperture with one receiver that measures only power of the incoming signal.

4.7.1 Array Topology

The antenna array is the key of the DF system. A small array with two-bit digital phase shifters and a processor that estimates bearing angle using only power measurements is presented [43]. Design is based on the holography principle [44], and a block diagram of the investigated antenna is shown in Fig. 4.11a. The linear antenna array consists of N equally half-wave spaced elements (below we use four elements as the base design for the experiments). Each of the $N - 1$ antenna elements with uniform amplitude distribution has electronically controlled two-bit digit phase shifters, for example, based on pin-diode switchers [45].

Two-bit phase shifter changes the phase of the signal captured by the antenna element in the range $0 \div 360^0$ with a discrete equal to 90^0. One reference channel without a phase shifter has a LNA with gain G. The antenna array has a simple summing feed network based, for example, on a microstrip printed parallel combiner. The output of the combiner is connected with a power estimator; the output of the power estimator is connected with a digital correlator; and the output of the digital correlator is connected with a beamforming processor that estimates the DOA.

It is known that the regular scanning phased array with two-bit digital phase shifters can significantly degrade beam pointing performance [42]. Consider, for example, an array with three omnidirectional elements (adjacent element distance is equal to half of the wavelength). Table 4.4 presents calculated two-bit phase values ϕ_n (in degrees) for each antenna element for scanning angle θ_0 in the range of 10° to 80° (Table 4.4).

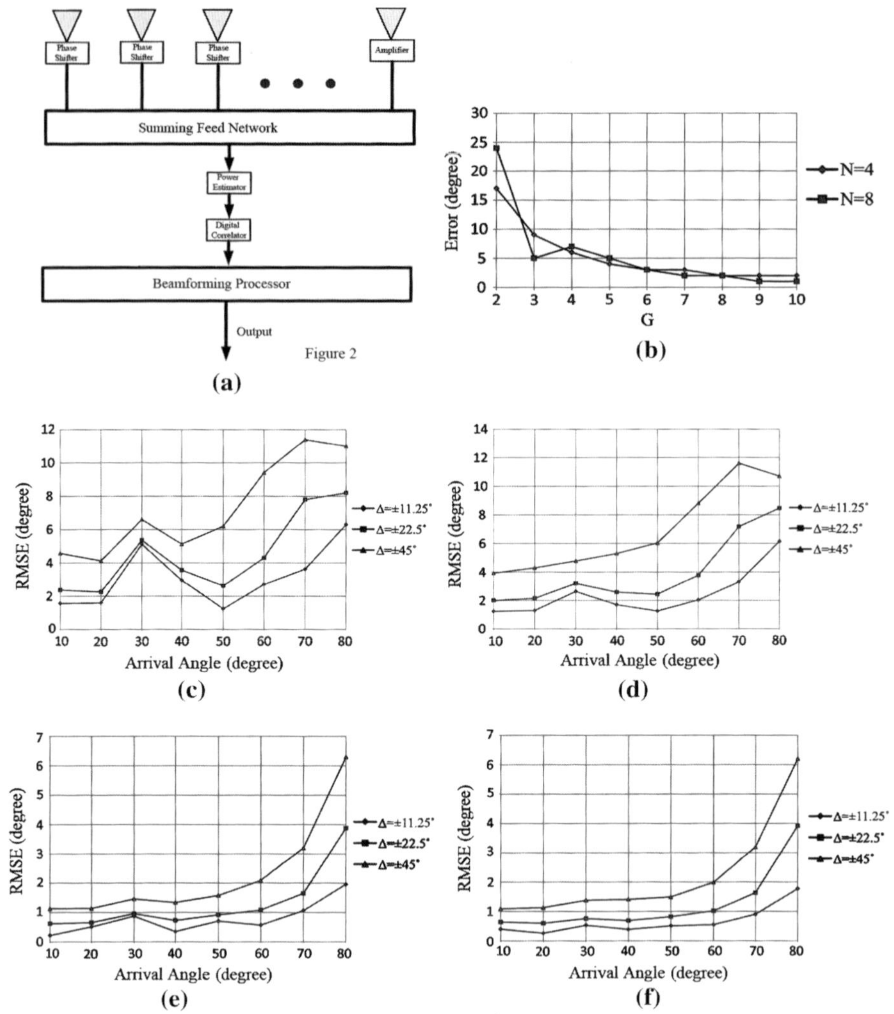

Fig. 4.11 Simulation results for small array with two-bit digital phase shifters and reference element **a** Array topology. **b** Maximum absolute angle error in beam pointing versus amplifier gain of reference element. **c–f** Root mean square error in beam pointing versus RF source location. **c** $N = 4$, $G = 15$ dB; **d** $N = 4$, $G = 20$ dB; **e** $N = 8$, $G = 15$ dB; **f** $N = 8$, $G = 20$ dB

As we see from the table, for scanning angles θ_0 from 10° to 20°, 50° to 60°, or 70° to 80°, two-bit phase shifters realize the same phase code combination. Hence, such phased array configuration does not provide accurate beam direction pointing. Presented solution consists of phase manipulation (using two-bit phase shifter) of the signal captured by each array element, recording of power variations received at the array output, processing of these variations with a digital correlator and beamforming processing that estimates DOA of the incoming signal.

Table 4.4 Phase shifts for regular three elements electrically beam scanning array with two-bit phase shifters within scanning angle range from 10° to 80°

θ_0	ϕ_1	ϕ_2	ϕ_3
10	0	90	90
20	0	90	90
30	0	90	180
40	0	90	270
50	0	180	270
60	0	180	270
70	0	180	0
80	0	180	0

4.7.2 Description of the Processing Algorithm

The proposed procedure includes measuring an antenna array power variations $P_k(\theta, r)$ (using power the estimator shown in Fig. 4.11a) caused by phase changes $\Delta\phi_r = \pi \cdot r/2$ of the kth element's phase shifter ($k = 1, 2, N - 1$). Angle position of the incoming source is θ.

Measured results are processed with a digital correlator

$$\mathrm{Re}_k(\theta) = \sum_{r=1}^{4} P_k(\theta, r) \cdot \cos(\pi \cdot r/2) \tag{4.22}$$

$$Jm_k(\theta) = \sum_{r=1}^{4} P_k(\theta, r) \cdot \sin(\pi \cdot r/2) \tag{4.23}$$

Values (4.22) and (4.23) can be expressed in complex form

$$E_k(\theta) = \mathrm{Re}_k(\theta) + j \cdot Jm_k(\theta) \tag{4.24}$$

or

$$E_k(\theta) = |E_k(\theta)| \cdot e^{j \cdot \psi_k(\theta)}$$

Calculated values in Eq. (4.24) are multiplied by complex weights $w_k(\theta_0) = e^{j\pi \cdot (k-1) \cdot \sin\theta_0}$ in the beamforming processor, and function $F(\theta, \theta_0)$ is estimated for different scanning angles θ_0

$$F(\theta, \theta_0) = \sum_{k=1}^{N-1} E_k(\theta) \cdot e^{j\pi \cdot (k-1) \cdot \sin\theta_0} \tag{4.25}$$

We will show that if an amplifier gain G of the reference channel is more than the number of the antenna array elements, then expression (4.25) determines the AOA of the RF incoming signal.

4.7.3 Algorithm Analysis

The power of the signal received by a linear phased antenna array (up to a constant factor independent of the element number) due to phase change at kth element in interval 0–360° with a discrete 90° can be expressed as:

$$P_k(\theta_0, r) = \left|B_k(\theta) \cdot e^{j\varphi_k} + D_k(\theta) \cdot e^{j(\pi(k-1)\cdot\sin\theta + \pi\cdot r/2)} + G \cdot D_N(\theta) \cdot e^{j(\pi\cdot(N-1)\cdot\sin\theta + \varepsilon)}\right|^2 \tag{4.26}$$

Array with equally half-wave spaced elements and uniform amplitude distribution consists of ($N - 1$) elements with phase shifters and one reference element with LNA; k = 1, 2, 3, ..., N−1.

$D_k(\theta)$ = radiation pattern of the kth element toward the arrival angle direction θ.

Reference element of the array is connected to a summing network via the LNA with a gain equal $G \cdot e^{j\varepsilon}$.

$$B_k(\theta) \cdot e^{j\varphi_k} = \sum_{n=1}^{N-1} D_n(\theta) \cdot e^{j\cdot\pi\cdot(n-1)\cdot\sin\theta} - D_k(\theta) \cdot e^{j\cdot\pi\cdot(k-1)\cdot\sin\theta} \tag{4.27}$$

Measured data (4.26) can be expressed as follows:

$$P_k(\theta, r) = S_{k0}(\theta) + S_{k1}(\theta, r) + S_{k2}(\theta, r) \tag{4.28}$$

where

$$S_{k0}(\theta) = B_k^2(\theta) + D_k^2(\theta) + (G \cdot D_N(\theta))^2 + 2B_k(\theta) \cdot G \cdot D_N(\theta) \cdot \cos(\delta_k) \tag{4.29}$$

$$\delta_k = \pi \cdot (N - 1) \cdot \sin\theta + \varepsilon - \varphi_k$$

$$S_{k1}(\theta, r) = 2B_k(\theta) \cdot D_k(\theta) \cdot \cos(\pi \cdot (k - 1) \cdot \sin\theta + \pi \cdot r/2 - \varphi_k) \tag{4.30}$$

$$S_{k2}(\theta, r) = 2G \cdot D_N(\theta) \cdot D_k(\theta) \cdot \cos(\pi \cdot (k - N) \cdot \sin\theta + \pi \cdot r/2 - \varepsilon) \tag{4.31}$$

For simplification, assume that radiation patterns of the different antenna elements $D_k(\theta)$ are identical, i.e., $D_k(\theta) = D(\theta)$; k = 1, 2, 3,...,N.

Substituting (4.28)–(4.31) into (4.22) and (4.23) the result can be written as:

$$E_k(\theta) = 4 \cdot D^2(\theta) \cdot e^{-j\cdot\pi\cdot(k-1)\cdot\sin\theta} \cdot \left(C_k(\theta) \cdot e^{j\varphi_k} + e^{j\cdot(\pi\cdot(N-1)\cdot\sin\theta + \varepsilon)}\right) \tag{4.32}$$

where $C_k(\theta) \cdot e^{j\varphi_k} = F_0(\theta) - e^{j\cdot\pi\cdot(k-1)\cdot\sin\theta}$, and $F_0(\theta) = \sum_{k=1}^{N-1} e^{j\cdot\pi\cdot(k-1)\cdot\sin\theta}$

Beamforming processor computes formula (4.25) based on the expression (4.32)

$$F(\theta, \theta_0) = \sum_{k=1}^{N-1} E_k(\theta) \cdot e^{-j\pi\cdot(k-1)\cdot\sin\theta_0} = F_1(\theta, \theta_0) + F_2(\theta, \theta_0) \tag{4.33}$$

where

$$F_1(\theta,\theta_0) = 4 \cdot D^2(\theta) \cdot \sum_{k=1}^{N-1} C_k(\theta) \cdot e^{j\varphi_k} \cdot e^{j\cdot\pi\cdot(k-1)\cdot(\sin\theta_0 - \sin\theta)} \tag{4.34}$$

$$F_2(\theta,\theta_0) = 4G \cdot D^2(\theta) \cdot e^{j\cdot(\pi\cdot(N-1)\cdot\sin\theta + \varepsilon)} \cdot \sum_{k=1}^{N-1} e^{j\cdot\pi\cdot(k-1)\cdot(\sin\theta_0 - \sin\theta)} \tag{4.35}$$

The output of the beamforming processor has two components: component (4.35), with a maximum absolute value $q_2 = 4G \cdot (N-1) \cdot D^2(\theta)$ directed to the scanning angle θ_0 equal arrival angle θ, and component (4.34), with absolute maximum value $q_1 = 4 \cdot D^2(\theta) \cdot |F_0(\theta)| \cdot (N-2)$ for $\theta_0 = \theta$. If the ratio q_2/q_1 is much larger than 1, then expression (4.33), as determined by (4.35), estimates AOA. Because the maximum value of $|F_0(\theta)|$ is equal to $|F_{0\,\max}(\theta = 0)| = (N-1)$, condition $q_2/q_1 \gg 1$ corresponds to the expression

$$\mathrm{G} \gg (N-2) \tag{4.36}$$

Ratio (4.36) means that amplifier gain G has to be much larger than the number of the elements. Correct AOA estimation as a function of the amplifier gain will be based on the simulation results of the proposed process.

4.7.4 Simulation Results

MATLAB software simulations are conducted for the equally spaced N omnidirectional antenna array system shown in Fig. 4.11a with uniform amplitude distribution and the distance among elements equal to half of the free space wave. Figure 4.11b presents the maximum absolute angle error in beam pointing as a function of the amplifier gain G. Each angle error is calculated as a maximum value within the arrival angle range θ of 5°–80°. It is seen that for the gain values 5 and more angle error does not exceed 5° for both four and eight element antenna arrays. It means that an acceptable DOA error can be reached for the amplifier gain value less than is determined by the ratio (4.36).

In practice, digital phase shifters exhibit discrete phase shifts $\pi \cdot r/2$ with errors $\xi_k(r)$, i. e. $\eta_k(r) = \pi \cdot r/2 + \xi_k(r)$ $(r = 0,\ 1,\ 2,\ 3)$, where $k =$ element number, $r =$ number of phase discrete. Typically, phase errors $\xi_k(r)$ vary randomly from one phase state to another and from one number of elements to the another. As a result, these errors cause additional antenna array errors in beam pointing, gain reductions, increasing of average SLL, perhaps shape change. For us, the most important parameter is the beam pointing value. Figure 4.11c–f shows simulation results of the root mean square error (RMSE) in beam pointing value (in degrees) as a function of the original RF source angle location. The RMSE results based on 500 random phase error sets $\xi_k^m(r)$ are drawn as a function of the arrival angle θ

(m = set number; n = array element number; r = 0, 1, 2, 3). Each phase error set values $\xi_k^m(r)$ are assumed to be uniformly and randomly distributed in the following intervals: $\Delta = \pm 11.25^0$ or $\Delta = \pm 22.5^0$, or $\Delta = \pm 45^0$ independent for different r and k numbers. Error $\varepsilon_m = \theta_{est}^m - \theta$ (m = 1, 2,...,500; θ_{est}^m = estimated arrival angle) is calculated for each m number and then RMSE $= \sqrt{\sum_{m=1}^{500} \varepsilon_m^2/500}$ is estimated. The results are presented for four and eight array elements. Simulation results show that the RMSE values calculated for two different amplifier gains equal to 15 dB and 20 dB are almost the same.

Multipath fading is another source of the error when estimating the beam pointing angle. Because the varying path lengths, phases, amplitudes, and angles of arrival for the reflected waves are random, the signal received by each array element becomes a random variable. The signal impinging on nth element of the equally half wavelength spaced antenna array with uniform amplitude distribution and identical radiation patterns of antenna elements can be expressed as:

$$S(n) = \alpha_0 \cdot D(\theta) \cdot e^{j\cdot(\pi\cdot(n-1)\cdot\sin\theta+\phi)} + D(\theta) \cdot \sum_{m=1}^{M} \alpha_m \cdot e^{i(\pi\cdot(n-1)\cdot\sin\theta_m+\gamma_m)} \tag{4.37}$$

The first component of the expression (4.37) is line-of-sight signal and second component includes a number of waves reflected and scattered by cars that surround driver's car. For computer simulation, we consider Jakes modeling method of scatters modeling [46]:

(a) Signals received by an antenna array are assumed to be plane waves;
(b) Random and independent, for different scattering waves, amplitudes α_m of scatters have Rayleigh distribution;
(c) Random phases γ_m are uniformly distributed over interval $[-\pi\,\pi]$ for all k and they are mutually independent;
(d) Angle of arrival θ_m is a random value uniformly distributed over the interval $\theta \pm \delta\theta$. Note, that θ is an angle of RF source arrival (amplitude of this signal is α_0). We used M = 50 harmonics for simulation process.

Figure 4.12 shows simulation results of the RMSE based on 500 random samplings for the signal to noise ratio 10 dB ($\sum_{m=1}^{M} \alpha_m^2/\alpha_0^2 = 10$). Calculated for four and eight elements array and for amplifier gain values 15 and 20 dB curves as a function of arrival angle θ are presented for different $\delta\theta$ values: ±10, ±20, and ±50°.

4.7.5 Experimental Results

Experimental DOA measurements were conducted with four elements array that include three elements with two-bit phase shifters and one element with LNA. Block diagram of the array is shown in Fig. 4.13a.

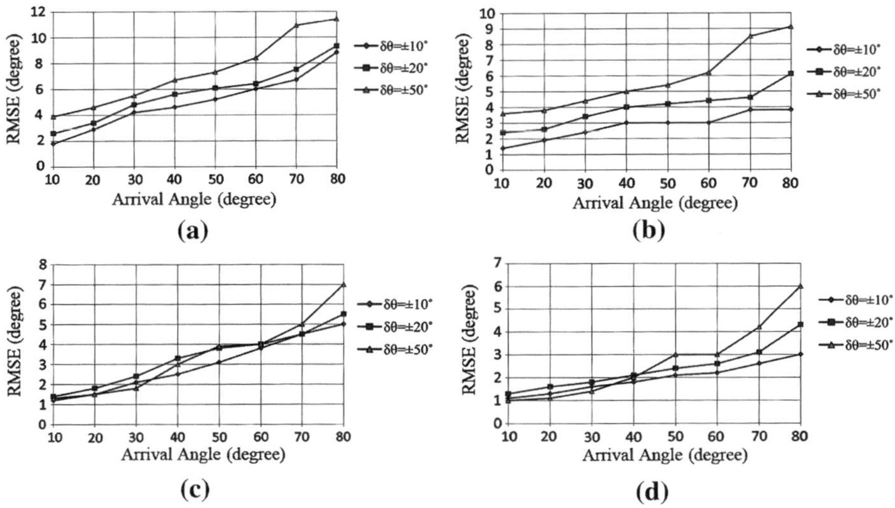

Fig. 4.12 RMSE for multipath fading as a function of versus RF source angle, results are based on 10,000 random sampling for SNR equal 10 dB. **a** $N = 4$, $G = 15$ dB. **b** $N = 4$, $G = 20$ dB. **c** $N = 8$, $G = 15$ dB. **d** $N = 8$, $G = 20$ dB

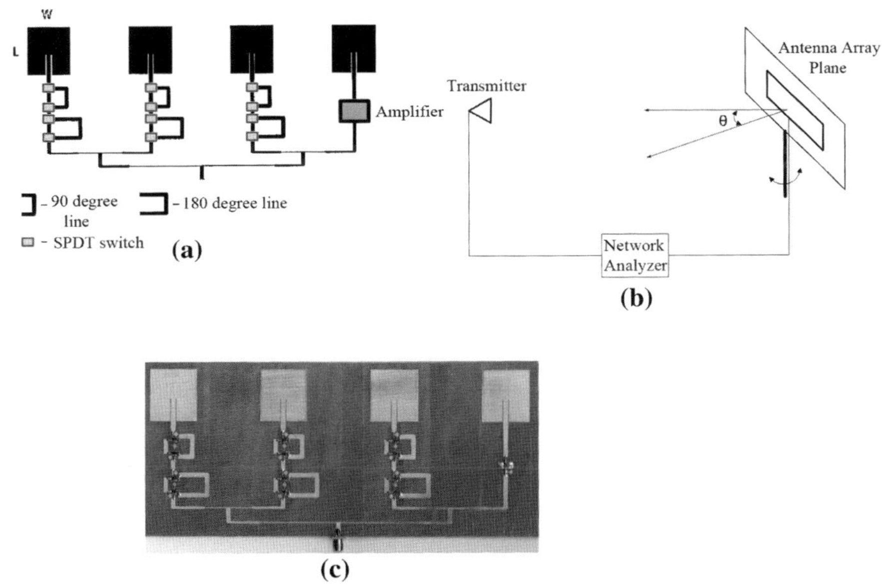

Fig. 4.13 Experimental set up for four linear array, three elements with two-bit digital phase shifters, one reference element with an amplifier. **a** Block diagram of the linear array. **b** Set up for measurements. **c** Antenna array prototype

Table 4.5 Power estimator measurements due to phase variations at the array Elements, angle of arrival $\theta = 20^0$

$k \backslash r$	r=0	r=1	r=2	r=3
k = 1	0	−0.2	2.29	1.58
k = 2	0	−1.6	0.31	1.39
k = 3	0	−2.76	−4.98	−0.42

Table 4.6 Power estimator measurements due to phase variations at the ARRAY elements, angle of arrival $\theta = 40^0$

$k \backslash r$	r=0	r=1	r=2	r=3
$k = 1$	0	−1.5	−3.8	−3.6
$k = 2$	0	3.3	4.5	0.3
$k = 3$	0	−4	0	−3.1

Each element operating at 2 GHz presents linear polarized patch antenna with the following dimensions: length $L = 35.9$ mm, width $W = 32$ mm. Inserted microstrip feed line section provides 50 ohm input element impedance. Two-bit phase shifters are implemented based on switched-line concept, which uses electronically controlled switches and lines of different lengths to obtain a set of phase steps. The outputs of the antenna array elements are connected with 50 ohm inputs of feed line that has 50 ohm output impedance and combines all signals into one output. AOA measurement setup is shown in Fig. 4.13b. The linear array mounted on turntable can change azimuth position relative to RF incoming source angle.

A photograph of the fabricated array prototype is shown in Fig. 4.13c. The designed array is fabricated on an $h = 1.6$ mm FR-4 substrate with an element separation equal to 7.5 cm which is 0.5 of the free space wavelength. The switch system consists of commercially available SPDT switches RF2436 from RF Micro Devices. The array is built according to the block diagram shown in Fig. 4.13a. Integrated amplifier ERA-6 (measured gain is about 9 dB) is used for amplification in the reference channel. The power supplier and control circuits are placed on the back side of the FR-4 board.

Two series of experiments were conducted with four antenna array elements: the first series refers to the RF wave coming from the angle $\theta = 20^{\circ}$ to the array normal and the other series to $\theta = 40^{\circ}$. Tables 4.5 and 4.6 show the power measurements results $P_k(\theta, r)$ (dB scale normalized to the value $P_k(\theta, 0)$ due to the differential phase shifts 0° ($r = 0$), 90° ($r = 1$), 180° ($r = 2$), and 270° ($r = 3$) realized by kth element phase shifter.

Using power measurement results and formulas (4.1)–(4.4), we estimated arrival angle θ_{est}: $\theta_{\text{est}}(\theta = 20^{\circ}) = 22^{\circ}$ and $\theta_{\text{est}}(\theta = 40^{\circ}) = 34^{\circ}$. Simulations based on an amplifier gain equal to 10 dB in the reference channel show $\theta_{\text{est0}}(\theta = 20^{\circ}) = 21^{\circ}$ and $\theta_{\text{est0}}(\theta = 40^{\circ}) = 36^{\circ}$. We see good agreement between the measurements and the simulation. The error 4° between the simulation and the

Table 4.7 DOA estimation (in degree) for random errors in power estimator measurements

±0.5 dB	35	36	38	35	37	38	38	35	36	38
±1 dB	35	34	36	34	39	34	35	34	37	37
±2 dB	33	30	34	26	42	44	45	35	34	38

original AOA for arrival angle equal to 40° can be reduced up to 1° by increasing the amplifier gain to the value 20 dB.

Table 4.7 demonstrates the effect of instrumental errors in power measurements. We used the power measurement data from Table 4.6, added random errors (imitating measurement errors) and performed DOA algorithm. Table 4.7 shows 10 random DOA estimations for uniformly and randomly distributed instrumental errors: row 1 corresponds to the interval ±0.5 dB; row 2—to the interval ±1 dB; and row 3—to the interval ± 2 dB.

It is seen that random errors in power measurements ±2 dB or less do not effect on DOA errors more than 10°.

References

1. Butler J (1965) Digital, matrix, and intermediate frequency scanning. Antennas and propagation society international symposium, pp 66–70
2. Allen B, Beach M (2004) On the analysis of switched beam antennas for the W-CDMA downlink. IEEE Trans Veh Technol 53(3):569–578
3. Kaifas T, Sahalos J (2006) On the design of a single-layer wideband Butler matrix for switched-beam UMTS system applications. IEEE Antennas Propag Mag 48(6):193–204
4. Yao J et al (2011) Microstrip branch-line couplers for crossover application. IEEE Trans Microw Theory Tech 59(1):87–92
5. Bhowmik W, Srivastava S (2010) Optimum design of a 4 × 4 planar Butler matrix array for WLAN application. J Telecommun 2(1):68–74
6. Siachalou E et al (2004) On the design of switched beam wideband base stations. IEEE Antennas Propag Mag 46(1):158–167
7. Gotsis K et al (2009) On the direction of arrival (DOA) estimation for switched beam antenna system using neural networks. IEEE Trans Antennas Propag 57(5):1399–1411
8. Qing X et al (2009) Sectored antenna array for indoor mono-station UWB positioning applications. Antennas and propagation, EuCAP 2009, pp 822–825
9. Karmakar N, Bialkowski M (2001) A beam forming network for a cellular switched-beam phased array antenna. IEEE Microw Wirel Compon Lett 11(1):7–9
10. Harrington R (1978) Reactively controlled directive arrays. IEEE Trans Antennas Propag 26(3):390–395
11. Schlub R, Lu J, Ohira T (2003) Seven element ground skirt monopole ESPAR antenna design using a genetic algorithm and the finite element method. IEEE Trans Antennas Propag 51(11):3033–3039
12. Taillefer E et al (2005) Direction-of-arrival estimation using radiation power pattern with an ESPAR antenna. IEEE Trans Antennas Propag 53(2):678–684
13. Ohira T (2006) Method for controlling array antenna equipped with a plurality of antenna elements, method for calculating signal to noise ratio of received signal, and method for adaptively controlling radio receiver, US Patent # 7,057,573, Publication Date June 2006

14. Taromaru M, Ohira T Electronically steerable parasitic array radiator antenna, principle, control theory and its applications. In: proceedings international union of radio science general assembly, New Delhi, India (October 2005)
15. Cheng J, Kamiya Y, Ohira T (2001) Adaptive beamforming of ESPAR antenna using sequential perturbation. In: Proceedings of IEEE MTT-S international microwave symposium digest, vol 1. May 2001, pp 133–136
16. Sun C, Hirata A, Ohira T, Karmakar NC (2004) Fast beamforming of electronically steerable parasitic array radiator antennas: theory and experiment. IEEE Trans Antennas Propag 52(7):1819–1832
17. Ozawa J et al (2010) Hamiltonian algorithm with momentum attenuation for adaptive beamforming of ESPAR antenna. Phased array systems and technology, IEEE international symposium, pp 990–994
18. Huang W et al. (2010) High-gain dual-band ESPAR antenna with simple on/off controlling. 9th international symposium on antennas, propagation and EM theory (ISAPE), pp 315–318
19. Milne R (1987) Adaptive antenna array, US Patent # 4,700,197, Date of Publication 1987
20. Sibille A et al (1997) Circular switched monopole arrays for beam steering wireless communications. Electron Lett 33(7):551–552
21. Frederik J et al (2002) A smart antenna receiver using a single RF channel and digital beamforming. IEEE Trans Microw Theory Tech 50(12):3052–3058
22. Goshi D et al (2004) Compact digital beamforming SMILE array for mobile communication. IEEE Trans Microw Theory Tech 52(12):2732–2738
23. Farzaneh S, Sebak A (2010) An optimum adaptive single port microwave beamformer based on array signal vector estimation. IEEE Trans Antennas Propag 58(3):738–746
24. See C (2003) A single channel approach to high resolution direction finding and beamforming. In: Proceedings of the IEEE international conference on acoustics, speech and signal processing, vol 5, pp 217–220
25. Tsui D (1996) Angle of arrival (AOA) solution using a single receiver, USA Patent 5497161
26. Goshi D et al A single RF channel smart antenna receiver array with digital beamforming, University of California, Electrical Engineering Department. http://www.ee.ucla.edu/~mwlab/poster/2004/Darren_ARR2004.pdf
27. Frederik J et al (2004) Smart antennas based on spatial multiplexing of local elements (smile) for mutual coupling reduction. IEEE Trans Antennas Propag 52(1):106–114
28. Kummer W et al (1963) Ultra-low sidelobes from time-modulated arrays. IEEE Trans Antennas Propag 11(6):633–639
29. Yang S et al (2003) Low sidelobe phased array antennas with time modulation. Antennas and propagation society international symposium. IEEE 4:200–203
30. Yang G, Nie Z (2010) Direction of arrival estimation in time modulated linear arrays with unidirectional phase center. IEEE Trans Antennas Propag 58(4):1105–1111
31. Li G et al (2009) A novel beam scanning technique in time modulated linear arrays. Antennas and propagation society international symposium. IEEE
32. Tong Y, Tennant A (2010) Low sidelobe level harmonic beam steering using time-modulated linear arrays. Antennas and propagation conference, Loughborough, pp 249–252
33. Li G et al. (2008) An adaptive beamforming in time modulated antenna arrays. International symposium on antennas, propagation and RM technologies, ISAPE, pp 166–169
34. Tong Y, Tennant A (2011) Beam steering and adaptive nulling of low sidelobe time-modulated linear array. In: Proceedings of the 5th European conference on antennas and propagation (EUCAP), pp 948–951
35. Jeon S et al (2002) A novel smart antenna system implementation for broad-band wireless communications. IEEE Trans Antennas Propag 50(5):600–606
36. Rocca P et al (2011) Synthesis of sub-arrayed time modulated linear arrays through a multi-stage approach. IEEE Trans Antennas Propag 59(9):3246–3254
37. Chen Y et al (2008) Adaptive nulling in time-modulated antenna arrays. International symposium on antennas, propagation and EM theory, ISAPE 2008, pp 713–716

38. Poli L et al (2011) Adaptive nulling in time modulate linear arrays with minimum losses. Technical report # DISI-11-102, University of Tronto, Italy. http://eprints.biblio.unitn.it/archive/00002095/01/DISI-11-102-R196.pdf
39. Haupt R (2006) Adaptive antenna arrays using a genetic algorithm, adaptive and learning systems. IEEE Mountain workshop, pp 249–254
40. Hansen C (2009) Phased antenna arrays, 2nd edn. Wiley, New York
41. Schmidt R (1986) Multiple emitter location and signal parameter estimation. IEEE Trans Antennas Propag 34(3):276–280
42. Hatcher B (1968) Granularity of beam positions in digital phased arrays. Proc IEEE 56(11):1795–1800
43. Rabinovich V (2011) Direction finding system for automotive applications using small phased antenna array. Microw Opt Technol Lett 53(10):2441–2446
44. Deschamps G (1967) Some remarks on radio frequency holography. Proc IEEE 55(4):570–571
45. Maloratsky L (2010) Electrically tunable switched line diode phase shifters part 1. High Frequency Electronics 9(4):16–23
46. Lee W (1998) Mobile communication engineering, 2nd edn. McGraw Hill, New York

Part II
Practical Design for Automotive

Chapter 5
Base Station Array Examples for Communication with Vehicles

Base station antenna arrays with different topologies are available on the market now. They provide wireless communication services such as WLAN, RFID, Intelligent Transportation Service (ITS), etc. A couple of excellent reference books [1, 2] describe the fundamental concepts and typical antenna engineering solutions for general base station applications. This discussion covers the design of the arrays that are more specific for automotive applications: toll collection systems, antenna arrays for vehicle identification at the parking lot area with high density of the cars, WLAN arrays with fixed beam, and configurations with electronically steerable beam. Main examples are related to the low cost and high efficient printed on circuit board antenna systems.

5.1 Fixed Beam Directional Arrays

5.1.1 Linear Polarized Toll Collection Design for 915 MHz

An overhead printed dipole array for electronic toll collection [3] is designed on the base of V-shaped printed dipole antenna element and operates within frequency a range 909–921 MHz. Single element structure, shown in Fig. 5.1a, consists of a V-shape printed dipole on the underside of the substrate and U-shape feed balun microstrip line on the upper surface of the substrate. The dipole is fabricated on the substrate RT4350 (dielectric constant is equal 3.35) with the thickness equal to 1.5 mm. Similar design of the printed dipole antenna element with an adjustable integrated balun for different frequency bands is presented in a number of papers [4, 5]. Figure 5.1b represents the sketch of the 4×6 array (4 rows of 6 elements).

Feed circuit is composed of four identical row feed boards which are connected to 4:1 combiner. The feed board shown in Fig. 5.1c is the 1 by 6 power divider. When designing the feed circuit, important requirement is co-phasing of the pass

V. Rabinovich and N. Alexandrov, *Antenna Arrays and Automotive Applications*,
DOI: 10.1007/978-1-4614-1074-4_5,

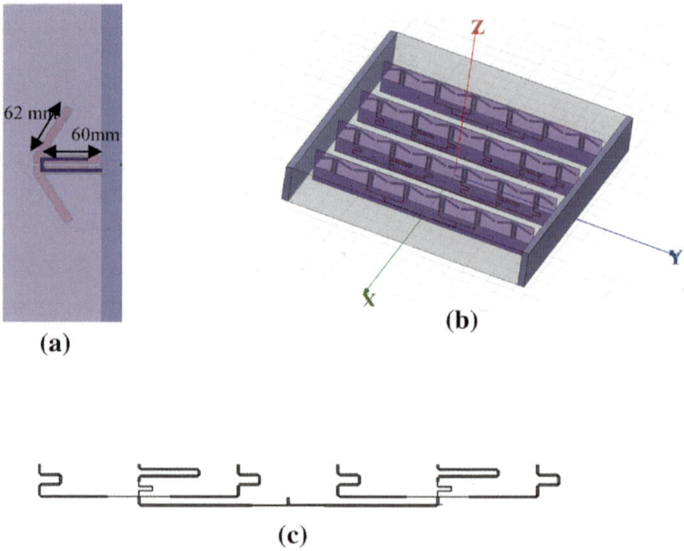

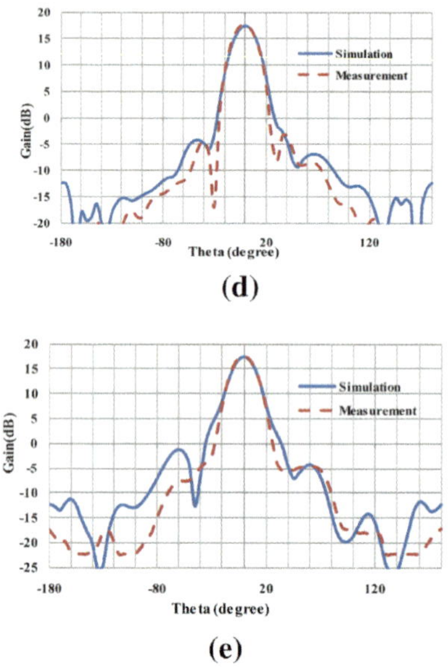

Frequency	909-921 MHz
Peak Gain	17 dBi
Sidelobe Level (SLL)	-20 dB
Return Loss	>10 dB
Polarization	Linear
3-dB Beamwidth E-Plane	18°
3-dB Beamwidth H-Plane	23°
Size of the Array (mm)	925×960×167
Dielectric Material, Thickness	RT4350, 1.52 mm

Fig. 5.1 **a** Individual element structure. **b** Sketch of 4 × 6 element array. **c** Feeding network topology. **d** E-plane radiation pattern. **e** H-plane radiation pattern, [3] © 2010 IEEE

lengths from each divider port element to the common output. Simulation shows a maximum phase difference equal 3°. Simulated and measured gain in E and H planes at 915 MHz are presented in Fig. 5.1d and e. It is seen that the peak gain is about of 17 dBi, half power (−3 dB) beamwidth for E and H radiation patterns is 18° and 23°, respectively, and −12 dB beamwidth is about 40° for both planes. Beamwidth 40° covers the road spot with 4 × 4 m if the array is mounted at a height of 6 m above the road. The numerical estimation of the power received by the transponder of the board unit (OBU), which is mounted on the car as the function of the road side unit (RSU) can be derived from Friis Transmission formula

$$P_r = \frac{P_t \cdot G_t \cdot G_r \cdot \lambda^2}{(4 \cdot \pi \cdot d)^2} \tag{5.1}$$

or in dB scale format

$$P_r = P_t + G_t + G_r + \text{Loss} \tag{5.2}$$

Assume that the transmitting power of the RSU is about $P_t \approx 15\,\text{dBm}$, the transmitting gain G_t corresponding to the angle directions ±20° from the normal to the array is $G_t = 5\,\text{dBi}$ (−12 dB from the maximum value equal 17 dBi), the gain of the receiving (OBU) antenna mounted in the interior part of the car is around $G_r \simeq -15\,\text{dBi}$, Friis Transmission Loss $= -20 \cdot \log(4 \cdot \pi \cdot d/\lambda) \approx -48\,\text{dB}$ for the frequency 900 MHz, and the distance between the transmitter and the receiver is equal d = 6 m. In this case, the received power calculated from (5.2) is equal $P_r \approx -43\,\text{dBm}$. This value is higher than the OBU sensitivity, which varies for different countries [6, 7] from −50 to −70 dBm. Power transmitted by transponder (uplink communication) is about of $P_t \approx 10\,\text{dBm}$, and the power received by RSU in this case is $P_r \approx -48\,\text{dBm}$, which is also higher than the RSU sensitivity equal −75 dBm. Hence, rough estimation of the link budget shows that the designed antenna array satisfies the ETC requirements. Table shown in Fig. 5.1 summarizes the measurement data obtained for the antenna array design presented in Fig. 5.1b. Figure 5.2 shows 900 MHz band transponder loop antenna printed on the circuit board in the left portion of the assembly.

Typically, transponder is attached to the car front window behind the rear-view mirror.

One of the known northeastern US E-ZPass electronic toll-collection system [8] uses reader antenna array, which is made of nine tuned plate elements in the 3 × 3 array operating at the 900 MHz frequency range. Each square element in the array has dimensions of about 150 × 150 mm and the total array size is about of 0.5 m.

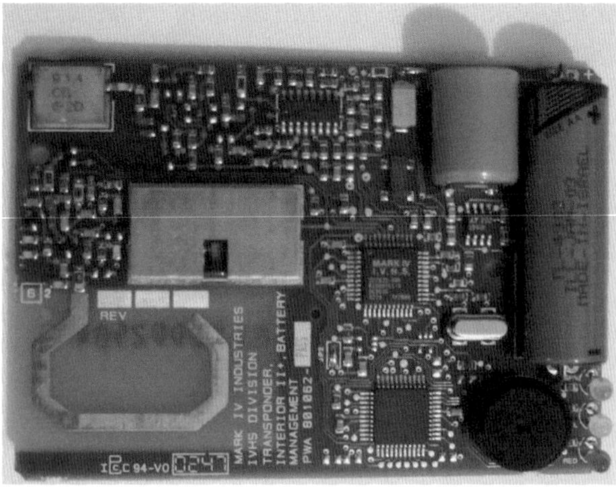

Fig. 5.2 Transponder loop antenna for ETC system

5.1.2 Circular Polarized 16-Element Microstrip Array for Toll Collection Application

According to China national specifications [7] ETC system has to operate in a 5.8 GHz frequency band. The main requirements for the reader antenna array parameters are as follows:

(1) Bandwidth: 5.78–5.85 GHz;
(2) Polarization: Right Hand circular (CP);
(3) Half power beamwidth: horizontal $<38°$ and vertical $<45°$;
(4) Gain: >14 dBi

The planar microstrip array [9] developed according to these requirements is presented in Fig. 5.3. Antenna array consisting of 16 patch elements operates at 5.795–5.815 and 5.855–5.875 GHz frequencies, and has right hand circular polarization (RHCP) design. Circular polarization of each square patch with linear dimensions equal $16.6 \times 16.6\ \text{mm}^2$ is obtained by the truncated corner with the length 2.6 mm. Adjacent patches are separated by distances of 21.49 and 21.52 mm in the X and Y directions, respectively. The antenna includes four subarrays (2×2 elements) with total dimensions 136.92×148.21 mm. The measured gain is more than 15 dBic, sidelobe level is about of -20, 3 dB beamwidth is 20°, and the axial ratio is <3 dB in the operating frequency band. Array is fabricated on a substrate with a 1.54 mm thickness and a relative dielectric constant of 2.17. It should be noted that this antenna array can be utilized not only for toll collection but for DSRC between a road side base station and a moving car system. A summary of data is presented in Fig. 5.3.

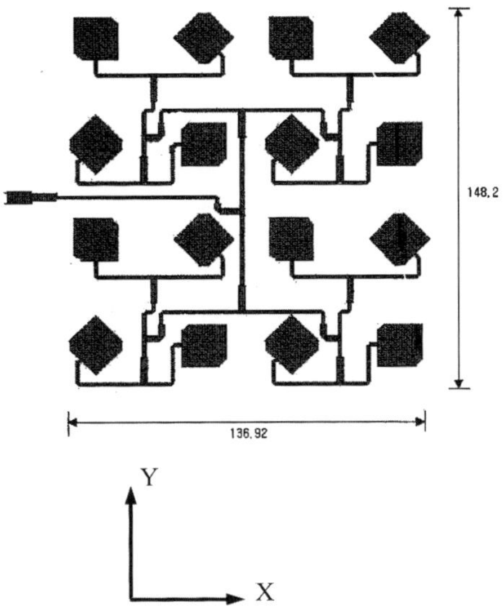

Frequency	5.795 GHz ~ 5.815 GHz 5.855 GHz ~ 5.875 GHz
Peak Gain	20.6 dBi
Sidelobe Level (SLL)	-20 dB
Return Loss	>10 dB
Polarization	RHCP
3-dB Beamwidth	20°
Size (mm)	196.4×148.2
Dielectric Material, Thickness, Dielectric Constant	Taconic TLY-5A-0620-C1/C1, 1.54 mm, ε=2.33

Fig. 5.3 Topology of planar 4 × 4 patch element array with circular polarization

5.1.3 Microstrip Array with Omnidirectional Radiation Pattern in Horizontal Plane for 2.45 GHz

Figure 5.4 shows the sketch of the low cost base station omnidirectional planar array. System can be used as drive-thru Internet antenna which provides Internet connectivity for some period of time as the vehicle passes through (or located in) the access point AP's coverage area. The array consists of the top and bottom traces [10].

The traces on the top layer of the antenna alternate from wide to narrow complementing the narrow to wide traces on the bottom layer. The last upper trace is narrow and shorted to the center of the last wide bottom trace. The length L of each section is approximately 0.275λ, the narrow line with width W_1 provides 50 ohm of sections is chosen such that it forms 50 ohm impedance of the microstrip line. The wide W_2 is approximately five times as wide as a narrow trace W_1. This antenna may be viewed as a set of $\lambda/2$ microstrip transmission lines. The total number of sections may be altered to provide a desired gain. Simulation analysis predicts a gain of 5.5 dBi for five sections, 6.9 dBi for seven sections, and 7.6 dBi for nine sections. Seven sections designed to operate at 2.45 GHz presented in Fig. 5.4 have the following dimensions: $W_1 = 2.06$, $W_2 = 16.25$ $L = 36.58$ mm. Shorting pins located on either end of the antenna have a 0.5 mm radius. The measured gain of the antenna shown in is about 5 dBi. Summarized data are presented in Fig. 5.4.

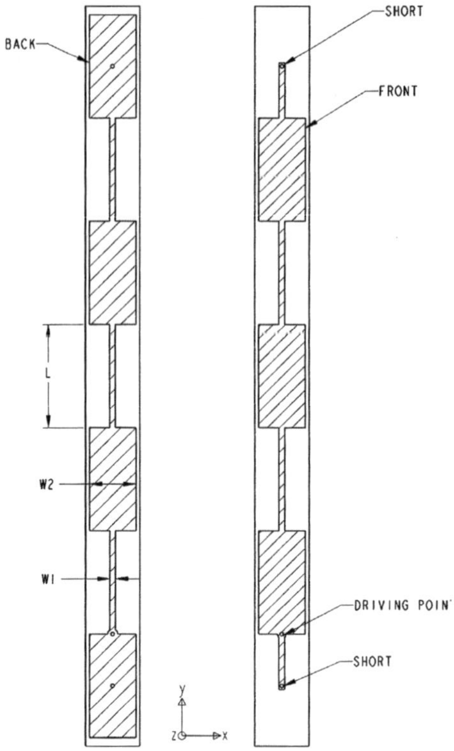

Frequency	2.4-2.48 GHz
Peak Gain	4.6 dBi
Sidelobe Level (SLL)	-11 dB
Return Loss	>10 dB
Polarization	Linear
3-dB Beamwidth	Omnidirectional in Horizontal Plane
Size for 7 Sections (mm)	260×20
Dielectric Material, Thickness, Dielectric Constant	Sheldahl ComClad, 0.762 mm, ε=2.6

Fig. 5.4 Printed omnidirectional planar array configuration

5.1.4 Printed Dipole Array with Reflector Plane for 2.45 GHz Band

A double side dipole array with a parallel reflector plane for 2.45 GHz shown in Fig. 5.5a (top and bottom sides) is one of the antenna systems manufactured by Canadian company SuperPass [11]. The patented antenna was developed by Waterloo University Electrical Engineering department. Each antenna element is formed by printing metal strip lines on both sides of the printed circuit board. Shorting inductive pin provides the operation in the required frequency band. The length of the single dipole arm L_1 is equal to 53.4 mm and the width W_1 is 3.15 mm. The reflector plane is mounted from the circuit board at a distance equal *to* $\lambda_0/4 \approx 25$ mm. A feeding network provides 50 ohm output impedance with return losses < -10 dB in the frequency range from 2 to 2.48 GHz. Summarized data for four element array are shown in Fig. 5.5b. A data summary table is presented in Fig. 5.5c.

A similar broadband double sided dipole array with a parallel reflector plane for 2.45 GHz [12] is demonstrated in Fig. 5.5b. The length of the single dipole arm

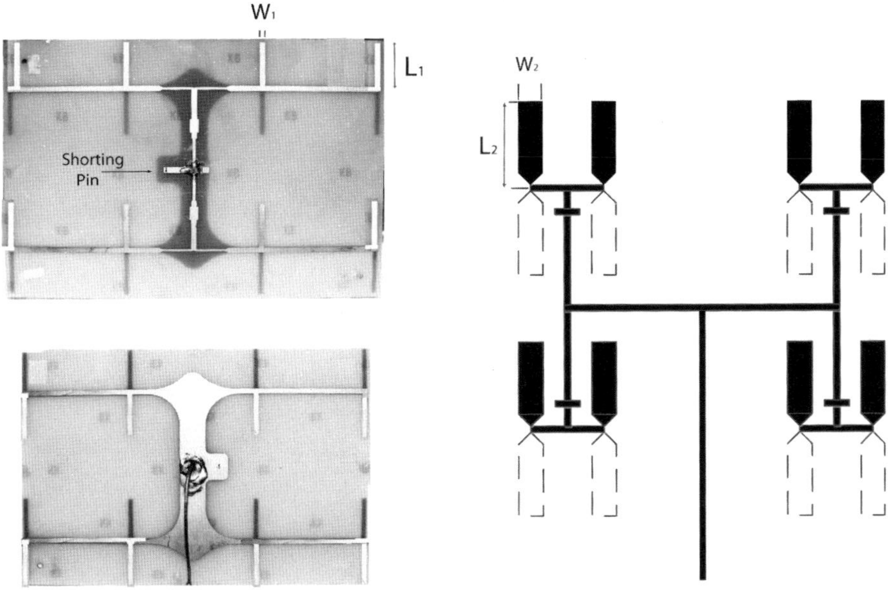

(a) (b)

Frequency	2.4-2.483 GHz
Peak Gain	14 dBi
Sidelobe Level (SLL)	<-11 dB
Return Loss	>14 dB
Polarization	Linear
3-dB Beamwidth E-Plane	27°
3-dB Beamwidth H-Plane	30°
Size (together with the case)	295×220×35

(c)

Frequency	2000-3000 MHz
Peak Gain	7 dBi
Return Loss	>10 dB
Polarization	Linear
3-dB Beamwidth	60°
Size	80×70
Dielectric Material	R04003 0.5 mm

(d)

Fig. 5.5 Double side dipole array with parallel reflector plane: **a** SuperPass design. **b** 4 × 4 2.45 GHz printed topology concept. **c** Data for superpass design. **d** Data for 2.45 GHz array

L_2 is equal to 26 mm and the width W_2 is 12 mm. The array prototype is fabricated on the substrate R04003 ($h = 0.5$ mm and dielectric constant 3.38). The reflector plane is mounted from the array at a distance equal to $\lambda_0/4 = 30$ mm. The total dimensions of the printed board are 80 × 70 mm^2. The feeding network provides 50 ohm output impedance with return losses < -10 dB in the frequency range from 2 to 3 GHz. Summarized data for four element array are shown at the table shown in Fig. 5.5d.

5.2 Antenna Arrays with Electronically Controlled Beam

5.2.1 System with ON/OFF Parasitic Microstrip Structure

Figure 5.6a shows the concept of the reconfigurable array [13]. The system consists of the horizontally polarized printed antenna element mounted at the center of the circle, as shown in Fig. 5.6a. The antenna is surrounded by N identical and equally spaced metallic parasitic radial sectors. As shown in Fig. 5.6b, parasitic elements can be electrically connected with pin-diode RF switches (ON position). Position OFF of the switching diodes corresponds to the disconnected radial elements. Connected radial sectors provide backward directional radiation, as shown by the arrow in Fig. 5.6b, and block the radiation in the opposite direction. An activated central antenna element is the antenna with fixed omnidirectional radiation pattern in the horizontal plane. The antenna system described in the reference paper [13] consists of 18 parasitic radial sectors and Z-shaped Alford loop central [14] printed antenna. F-type topology presented in Fig. 5.6d is also good candidate for operating in such systems. The array, described in reference paper [13] with $N = 18$ parasitic elements for 2.45 GHz application, is fabricated and tested for the following geometry: the outer diameter of the antenna (with parasitic elements) is equal to 72 mm, the inner diameter of the area, which occupies Z-shape antenna element, is equal to 28 mm. Geometry of Alfred loop printed on FR4 substrate (thickness is 0.8 mm and dielectric constant 4.6) (Fig. 5.6c), has the following strip line dimensions: $l_w = 17.8$, $W_w = 1.5$, $W_a = 6.36$ mm.

Radial angle, which occupies single parasitic element, is equal to 17°, and the angle gap among the adjacent radial sections is 3°. It was found, that the minimum number of connected parasitic sectors to obtain a minimum front to back ratio more than 5 dB and minimum VSWR < 2:1, is equal to 3. The measured VSWR values are < 2.5:1 within the frequency range from 2.3 to 2.6 GHz. Experimental results show that connection of the parasitic elements with numbers 13, 14, 15, 16, 17, and 18 leads to the array main beam directed to $\phi \approx 345°$; when the radial elements with numbers 12, 13, and 14 are connected, the angle direction of the main beam is $\phi \approx 27°$, for connections of the element numbers 2, 3, 4, 5, 6, and 7, main beam angle direction is equal to 165°. The beamwidth of the antenna system in the horizontal plane for horizontal wave polarization is about 120°.

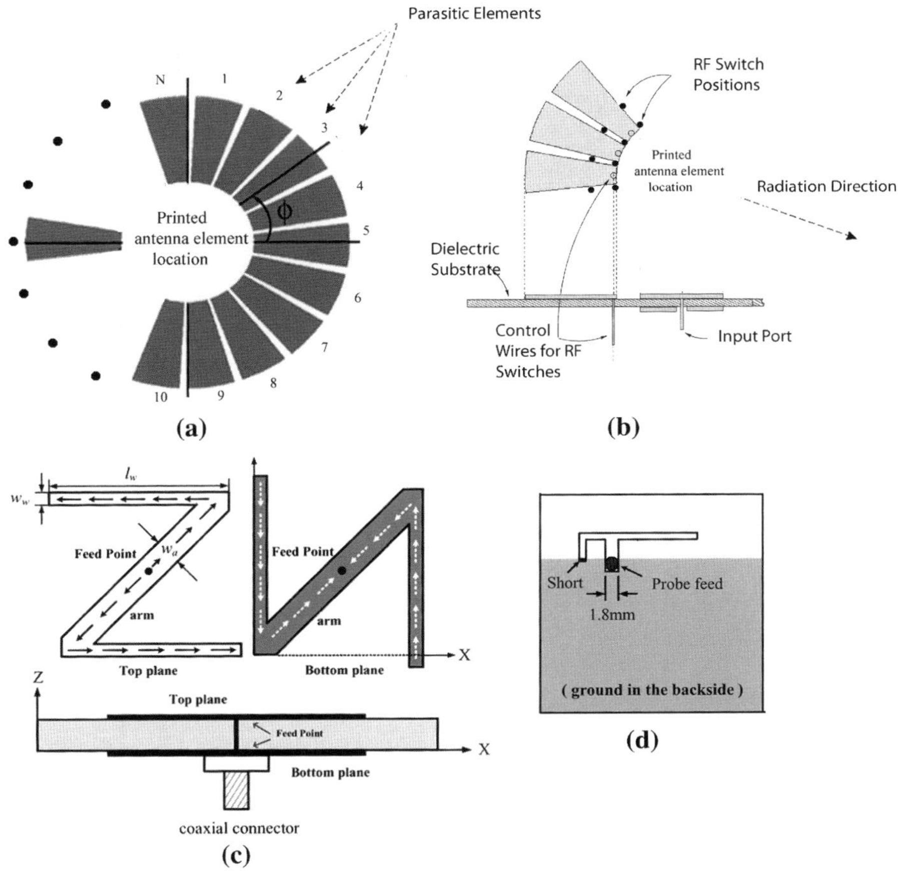

Fig. 5.6 Array with ON/OFF parasitic microstrip Structure: **a** Horizontally polarized concept structure. **b** RF switch positions and direction of radiation. **c** and **d** Printed antenna element configurations

5.2.2 Array for Horizontal Over 360° Beam Steering with Switchable Printed Landstorfer/Yagi

This section presents a switchable antenna array prototype that provides a 360° (in azimuth) electronically steerable beam. The array consists of six linear polarized directive antenna elements: three Landstorfer antennas (Fig. 5.7a) and three Yagi elements (Fig. 5.7b) [15]. The antenna elements are fabricated on a Roger 4003 substrate PCB with the thickness of 0.5 mm. The topology of the feeding network with Yagi, Landstorfer antenna elements, and a pin-diode switch system is shown in Fig. 5.7c. The pin-diode system, marked by letter S, operates in open/short

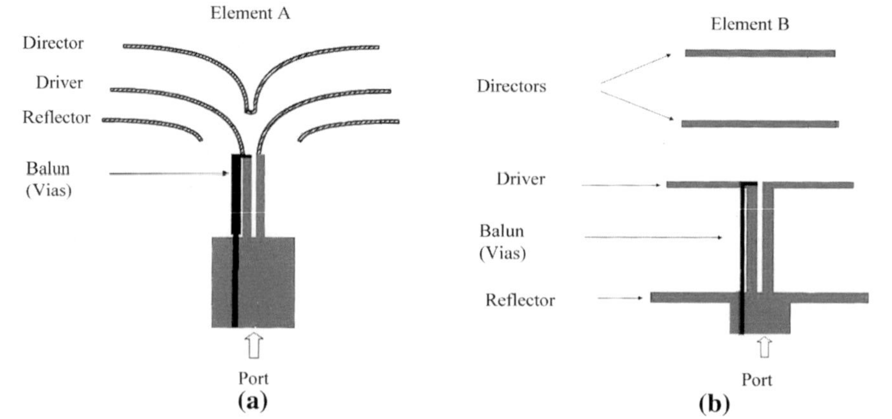

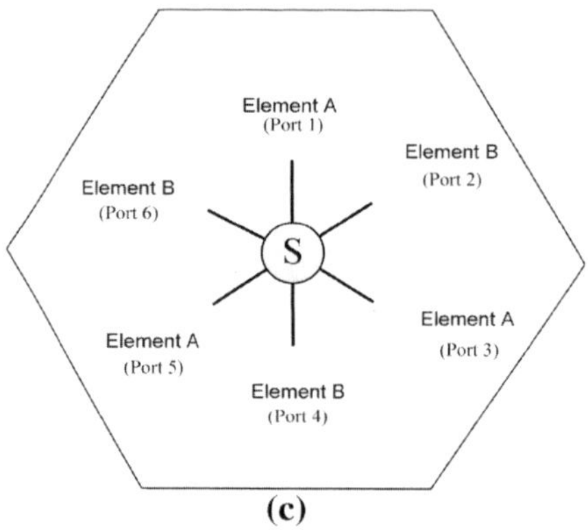

Frequency	2.4-2.48 GHz
Peak Gain Landstorfer	9.5 dBi (8.6 dBi)
Peak Gain Yagi	7.5 dBi (5.6dBi)
Return Loss	>10 dB
Polarization	Linear
3-dB Beamwidth Yagi/Landstorfer, single element E-Plane	59.5°/36°
3-dB Beamwidth Yagi/Landstorfer, single element H-Plane	88°/89°
Size (mm)	269×269
Dielectric Material, thickness	Roger 4003, 0.5 mm

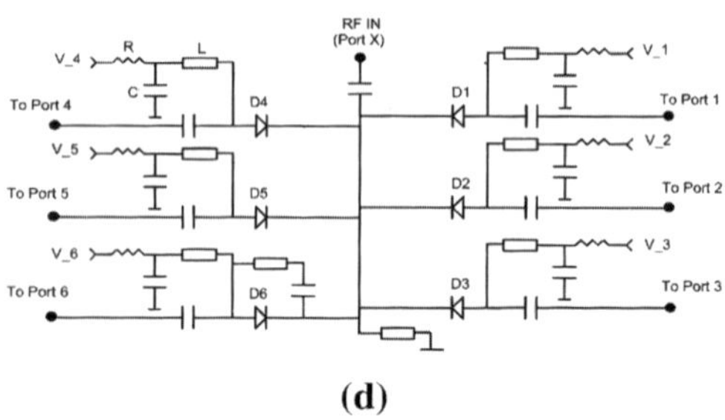

Fig. 5.7 Landstorfer/Yagi array: **a** Landstorfer element. **b** Yagi element. **c** Feeding network concept. **d** Simplified control circuit topology

(OFF–ON) states and provides scanning beam over 360° in azimuth. The topology of the feeding network with three Yagi and three Landstorfer elements provides smaller dimensions in comparison with six Landstorfer elements and greater gain in comparison with six Yagi elements. The overall size is 269 mm (269 mm corner to corner) with hexagonal shape. The array contains six ports: Port1, Port3, Port5 are the Landorsdorfer element ports associated with the Landstorfer antennas. The other three ports: Port2, Port4, Port6 are related to the Yagi antenna elements.

Each of six antenna elements can be selected to orient an independent directive beam using switching network shown in Fig. 5.7d. The RF input port and output ports are labeled by RF IN (Port X) and Ports 1–6, respectively. While one of the control ports (V 1 to V 6) is triggered by a DC voltage (3.3 V), the corresponding path is turned on. For example, when the PIN diode (D1) is turned on, the RF signal can go through from RF IN (Port X) to Port 1. The RF IN (Port X) is isolated from the other Ports (i.e., Port 2, 3, 4, 5, 6) since the PIN diodes D2, D3, D4, D5, D6 are off. Each antenna element is activated by applying 3.3 V on its control pins, respectively. Summarized data, obtained by testing the antenna prototype, are shown in the Table of Fig. 5.7.

5.2.3 Example of Practical Array with Simple 4 × 4 Butler Matrix

Figure 5.8a presents a practical example of four patch element antenna array [16, 17] with etched on the same board 4 inputs/4 outputs Butler Matrix. All four radiating elements of the antenna array are identical, and they are designed by utilizing rectangular patch shape. Each antenna element of the array operates within the frequency band of 5.25 GHz. The effective length and width of the patch antennas are calculated and optimized using simulation Ansoft HFSS software. A Butler matrix with the antenna elements are fabricated on RT/Duroid 5880 substrate, which has dielectric constant $\varepsilon_r = 2.2$, dissipation factor tan $\delta = 0.0002$, and thickness of 0.787 mm. The Butler matrix includes 3-dB directional couplers (quadrature hybrid couplers), two fixed strip line 45° phase shifters, and two crossovers, which provide isolation among signals at the crossing of lines. Quadrature hybrid is 3-dB directional coupler with 90° phase difference in the outputs of the through and coupled arms.

Basic operation of coupler is as follows. With all ports matched, the power entering port P_1 is divided between ports P_2 and P_3, with a 90° phase shift between them. No power is coupled to port P_4. Crossover is an efficient circuit, which provides two transmission lines crossing with a minimal coupling between them.

Table 5.1 shows a summary of the corresponding magnitudes and phase shifts between the inputs and the outputs of the matrix.

The array exhibits four directional beams in the different angle directions as shown in Fig. 5.8. Figure 5.8b, c, d, and e illustrate radiation patterns of 1R, 2L, 2R, and 1L beams, respectively. It is seen from the measurement results that the angle positions of the beams are as following: 16°, −39°, 38°, and 15°. For the beams at −15° and 16° the side lobe level is lower than −10 dB, however, for 2L and 2R beams SSL is greater than −10 dB.

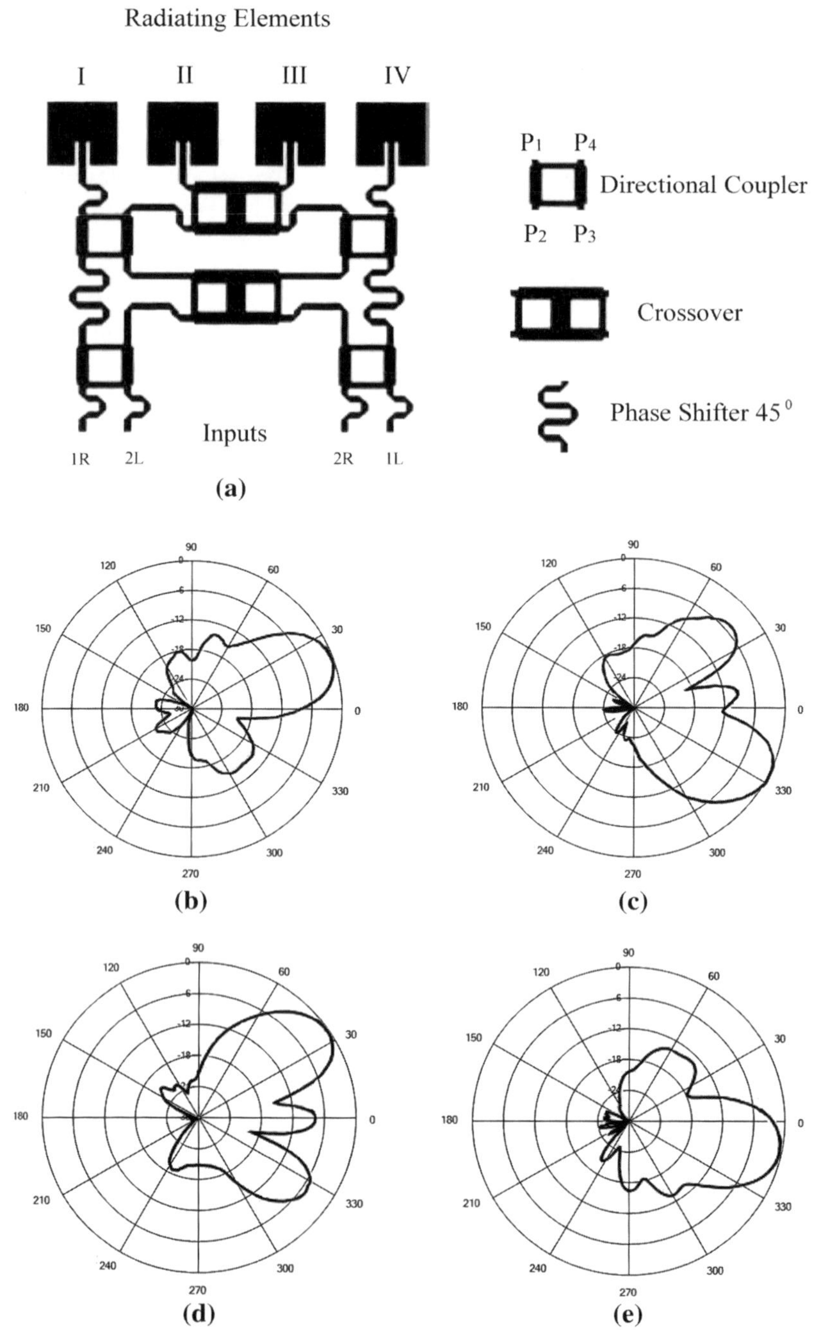

Fig. 5.8 Linear four element array with Butler matrix: **a** Array layout. **b** Radiation pattern from 1R output. **c** Radiation pattern from 2L output. **d** Radiation pattern from 2R output. **e** Radiation pattern from 1L output

Table 5.1 Magnitudes and phase shifts at the inputs and outputs of Butler matrix

Port	1R	2L	2R	1L	I	II	III	IV
Mag	1	0	0	0	0.524	0.506	0.472	0.48
Phase	0	0	0	0	129.4	90.8	47.98	−1.45
Mag	0	1	0	0	0.481	0.496	0.507	0.497
Phase	0	0	0	0	44.94	175.3	−54.4	88.75
Mag	0	0	1	0	0.497	0.507	0.496	0.481
Phase	0	0	0	0	88.75	−54.4	175.3	44.94
Mag	0	0	0	1	0.48	0.472	0.506	0.524
Phase	0	0	0	0	−1.45	47.98	90.83	129.4

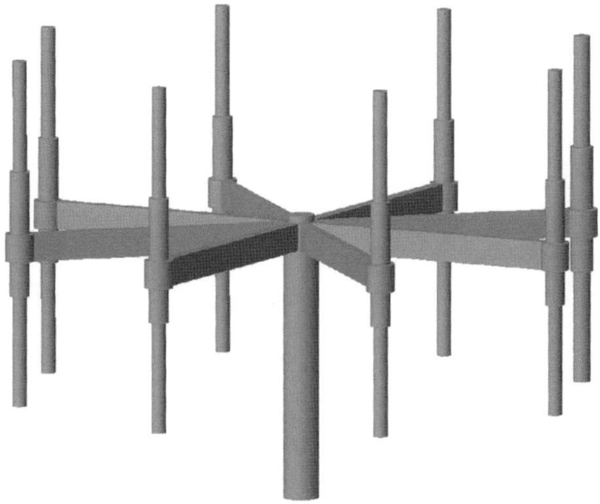

Fig. 5.9 Circular array with dipole elements

5.2.4 Circular Array with Electronically Controllable Dipole Elements

Commercially available circular base station array [18] with vertically oriented dipole elements is presented in Fig. 5.9.

The antenna consists of eight half-wave dipoles mounted on the circle with the diameter equal to operating wave length. The distance among neighboring dipoles is about 0.4λ. The feeding network with electronically controllable phase shifters and pin-diode attenuators/switches supports a few different operation modes:

1. Simple co-phasing of the signal from all elements in the predetermined azimuth angle position
2. Diversity (MIMO) operation with a few antenna elements
3. Adaptive operation with suppression side lobe suppression or even deep nulling in any user direction, when adjusting phases of each dipole in the array.

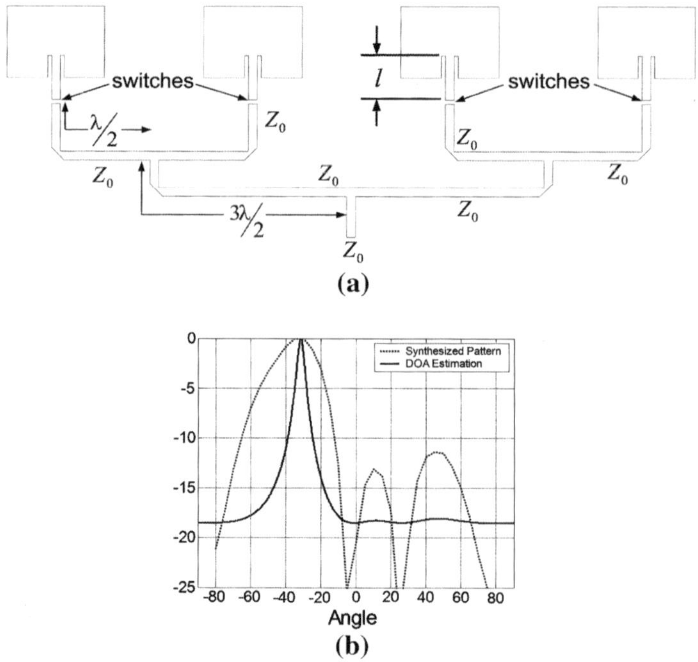

Fig. 5.10 **a** Concept of linear four element patch array with "ON/OFF" switches. **b** MUSIC (DOA Estimation) algorithm and Fourier (synthesized pattern) method

For conventional regime (without adaptation) measurements show gain of about 7.5 dBi in azimuth plane, sidelobe equal to −12 dB, and −3 dB beamwidth of about 80°. Adaptation toward the interference signal leads to the side lobe suppression of about −25 dB.

5.2.5 Single RF Channel Digital SMILE Array

Four element patch antenna array example, presented in Fig. 5.10a, is used for demonstration of a single receiver technique [19–22].

The array spacing is half of free-space wavelength at the resonance frequency of 5.8 GHz. The patch antennas with feed network are etched on a 1.57 mm Duroid substrate with dielectric constant equal to 2.33. Because each antenna element is switched ON/OFF, the loading of the antenna elements to the feeding network changes from 50 ohm to the open circuit. In this design, all transmission lines have the same characteristic impedance Z_0 with unbalanced T-junctions, as shown in Fig. 5.10a. The path lengths of the feed network are multiple of $\lambda/2$, and, therefore, only one path matched to Z_0 will be seen by the signal. It was found [19] that the length l from the inset feed of the element to pin-diode location can significantly affect the radiation pattern of the individual element. The pin-diode location can change the variations of the radiation pattern values up to 7 dB [21]. The element radiation pattern for the best location of the pin-diode is almost identical to the radiation pattern of the isolated element. After the signals received by each channel are recovered, the DBF processor estimates the direction of arrival (DOA) using MUSIC technique and regular Fourier method, as shown in Fig. 5.10b. Error for DOA estimation with MUSIC method does not exceed a few degrees. The array uses HJFET LNA elements NE32485C as the switching devices.

5.3 Parking Lot Vehicle Localization Antenna Systems

Vehicle localization is the part of ITS, which includes different aspects of traffic control managing. The difficulty of finding a car in a large or multilevel parking lot, at shopping or airport centers, close to the music or sporting events can be very annoying. The problem is so common that many parking ramps have special cars to drive people around to help them find their car. For example, GPS system [23] can be used in the open parking area. However, GPS signal strength is not enough to locate the vehicle in closed parking lots. Cost is another issue because such technique requires GPS equipment in each car that uses DF service.

According to the logic of the simplest DF system with an omnidirectional vehicle mounted antenna [24] car owner presses a button of the hand-held transmitting module, and a receiver, mounted on the vehicle, turns on a flushing light or horn sound, which can be recognized by the motorist. However, the area in the big parking lot includes a lot of sections and car owner can miss a response from the light or sound signal.

More sophisticated and advanced DF systems have an antenna array mounted on each vehicle, which is equipped with this service [25–31]. The antenna array elements can be combined using different schematics. Low cost DF system solution assumes a single receiver with multiple array elements and some kind of switching logic among the elements. It is known [25, 31] two different communication scenarios that are utilized for the localization of the vehicle at the areas, where visual finding of the car is difficult due to the high density of the parked vehicles. We discuss both scenarios: first, based on mounting antenna array on each vehicle, and second that uses single antenna array mounted on the lighting pole at the parking area.

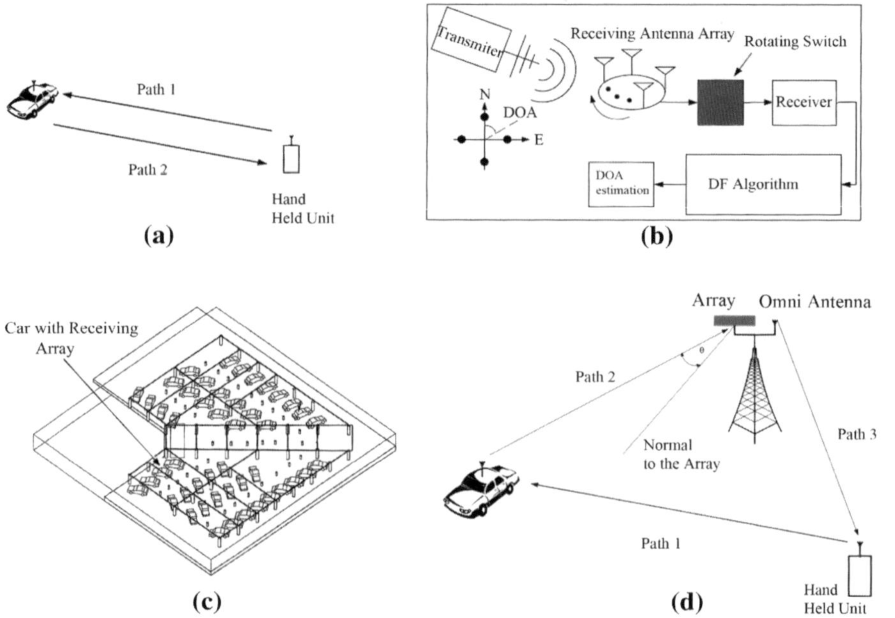

Fig. 5.11 **a** Communication scenario for the array mounted on the car. **b** Circular array with single receiver for DOA estimation. **c** Simulation test set up for vehicle localization in the car garage. **d** Communication scenario with antenna array positioned on the lighting pole

5.3.1 Communication Scenarios

The simplest communication scenario is demonstrated in Fig. 5.11a. RF signal radiated by an antenna of the hand-held car owner transmitter/receiver unit (Path 1) is received by an antenna array mounted on the car. Antenna array, mounted on each vehicle, includes single receiver and the electronically controllable switch connecting one array element with a receiver at one time, as it is shown in Fig. 5.11b. The antenna elements of the array are arranged circumferentially around their reference point, spaced apart at equal angles, for example 45°, when using eight array elements. Each directional antenna element covers the certain predetermined angle sector so that the total covered area of the system is 360° in azimuth. Each antenna is connected to the rotating switch that completes the revolution during 1/8 of the dwell time. The signal from the antenna passes through a single digital receiver, digital algorithm compares the power levels received by each antenna element, sorts them according to the signal strength, and the angle of the incoming signal corresponds to the angle sector with the highest received signal. The signal with DOA information is transmitted (using small omnidirectional antenna) back (path 2) to the car owner and activates, for example, car owner compass [32], which shows the car angle location. Note that the system requires the mounting of the antenna array on the each car equipped with this service.

Figure 5.11d shows an advanced communication scenario. The RF signal radiated by an antenna of the hand-held car owner transmitter/receiver unit (Path 1 shown in Fig. 5.11d) is received by a compact omnidirectional antenna mounted on the car. For example, car roof omnidirectional antenna, which described in reference paper [33], provides a communication range of about 200 m.

The omnidirectional antenna of the car retransmits the signal (Path 2 in Fig. 5.11d) with identification code that belongs to the owner vehicle. Retransmitted signal is received by the antenna array system mounted on the parking lot base station tower. Note that the array processor is activated only by the signals with the identification code, which specifies signal transmitted by the car (not hand-held unit). The processor calculates DOA of the vehicle with respect to the reference angle position, and omnidirectional antenna of the base station transmits signal with DOA data (Path 3) toward the driver's hand-held unit. As a result, car owner obtains car location information. Direction information can be updated as car owner moves relative to the car. It is very important that, according to the second scenario communication, the system uses only one sophisticated antenna array, which is installed on the special beacon or on the lighting pole at the parking area.

5.3.2 Single Channel Eight Element Circular Array for 2.4 GHz Applications

This section presents simulation results of the vehicle localization at the parking garage area with circle array, which has eight directional antenna elements [25]. Parking garage is the area with dimensions of 60 × 60 m and 46 vehicles surrounding a testing vehicle, which has a receiving antenna array. Each car has the following size: 4.8 × 1.8 × 1.5 m^3. The testing car is located in environment with multiple reflections, which are modeled using commercial software package Wireless InSite [34]. The circular array, consisting of eight equally spaced antenna elements, is centered on the vehicle's roof at the height 1.5 m. Array radius is equal to quarter wave length. Total signal captured by the individual antenna element is modeled as a combination of the direct signal, reflected and diffracted signals from the surrounded objects, such as cars, celling, floor, support spots, and so on. The floor, celling, and support posts are comprised of concrete, entry way massive room is modeled as a cube made of concrete, and the vehicles are made from the perfect electric metal body and rubber wheels (dielectric constant is equal to 3). More detail description of parking garage, channel model, and the propagation scenario is given in [34].

Table 5.2 Pass rate percentage for the highest RSSI algorithm

	Beamwidth 45°	Beamwidth 90°
Height 1.0 m	64.8	51.6
Height 1.25 m	70.5	52.6
Height 1.50 m	82.8	68.0
Height 1.75 m	84.8	59.8

The simulation is performed for the antenna element radiation pattern, which covers sector of 45 or 90° in azimuth in both E and H planes. The antenna array is connected to the rotating switch, as shown in Fig. 5.11b, that completes a full revolution (360°) during the time interval equal *to* T_{360}. The connection time with each array element is equals to $T_{360}/8$. The single receiver connected with each antenna element estimates a power received by the different antenna elements, compares these values, and chooses the angle sector with highest received signal strength indicator (RSSI) power level. This sector is determined as the DOA estimated sector. Sets of 122 different transmitter possible positions are placed on the horizontal plane of the parking garage area to provide statistical confidence in the evaluation of the described algorithm. Each transmit point can be thought of as user at the specific location to obtain a unique DOA estimate. The transmitter has a vertically polarized dipole antenna operating at 2.4 GHz within bandwidth of 1 MHz. Figure 5.11c shows a garage configuration used in simulation scenario and the position of the car with the receiving antenna array. Table 5.2 shows pass rate percentage for the RSSI algorithm in terms of the four different heights of the transmitting dipole antenna. Pass rate percentage is defined as the total number that falls within ±22.5° of the true DOA out of the possible 122 points.

It is seen from the table that the receiving antenna with 45° beamwidth exhibits better results. This is because the antenna with greater beamwidth, and, therefore, smaller gain has an increased beam pointing error due to the multiple reflections. Also, it can be noted that the results for greater antenna heights are better due to the lower level of reflections from the floor. Algorithm performs best, when the receiving antenna array is placed on the vehicle's roof.

5.3.3 Example of Conformal Smart 2.45 GHz Array for Lighting Pole Installation at the Parking Area

Lighting pole-mounted base station antenna [35–37], shown in Fig. 5.12a, allows estimating the azimuthal angle location of the radiating objects (in comparison with the reference position). Array consisting of six pentagonal elements covers 360° in azimuth. The array with diversity mechanism enabling the selection one of the six radiating array elements, is capable of operation in either right or left circular polarization. Common topology of the individual element and the array is shown in Fig. 5.12b and c, respectively. The array element is a slit disk antenna

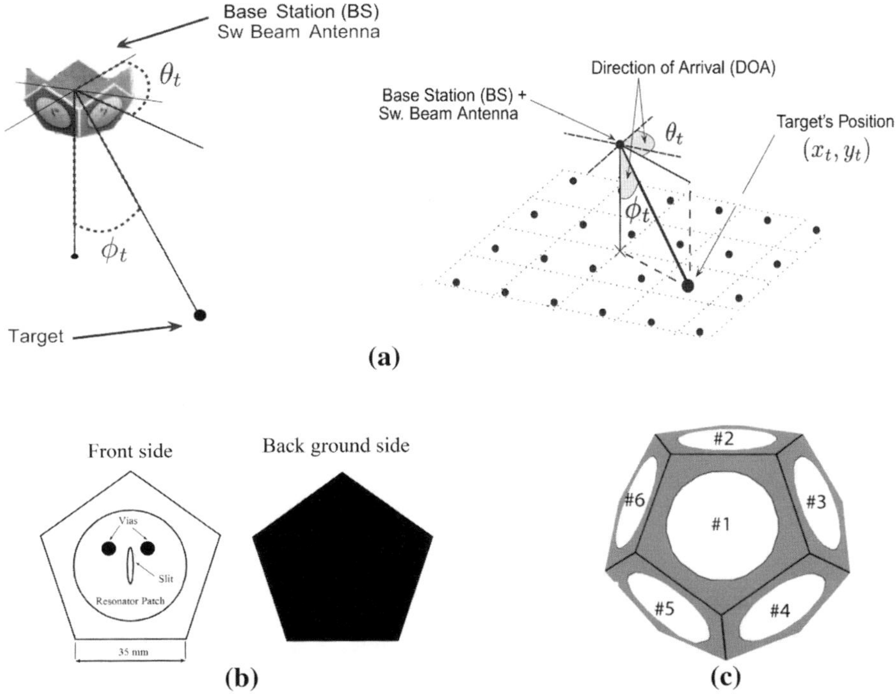

Fig. 5.12 **a** Angle location of radiating object. **b** Element concept of the array element. **c** Antenna array topology

etched on FR4 material with 1.6 mm thickness on the pentagonal ground plane. The ground plane of each element is taken only slightly larger than the circular patch resonator dimensions (Patch diameter is 33.5 mm). The antenna can operate on circular polarization by appropriate placement of the probe position.

Figure 5.13 presents the concept radiation patterns which show the maximum radiation angle for different antenna elements (Fig. 5.12c). Directions of the LHCP radiation pattern maximum as a function of the angle θ are shown by the arrow for each of all five antenna element locations (excluding element #1). Almost omni-directional radiation pattern of the antenna #1 is shown in Fig. 5.13 as a black circle spot. The gain level of this pattern is around -12 dB in comparison with the maximum radiation patterns measured for the remaining five antennas.

The beamwidth of the radiation pattern for each of the antenna elements (#2 to #6) is between 50° and 60°. This design allows using different processing algorithms for DOA estimation. Originally, the system was suggested to estimate angle of arrivals for the indoor radiating objects: the achievable accuracy was about of a few meters when using high resolution MUSIC algorithm. However, presumably, this system can be utilized for OAA estimation of the car in the big parking lot area where it is necessary to provide beam steering of 360° azimuth.

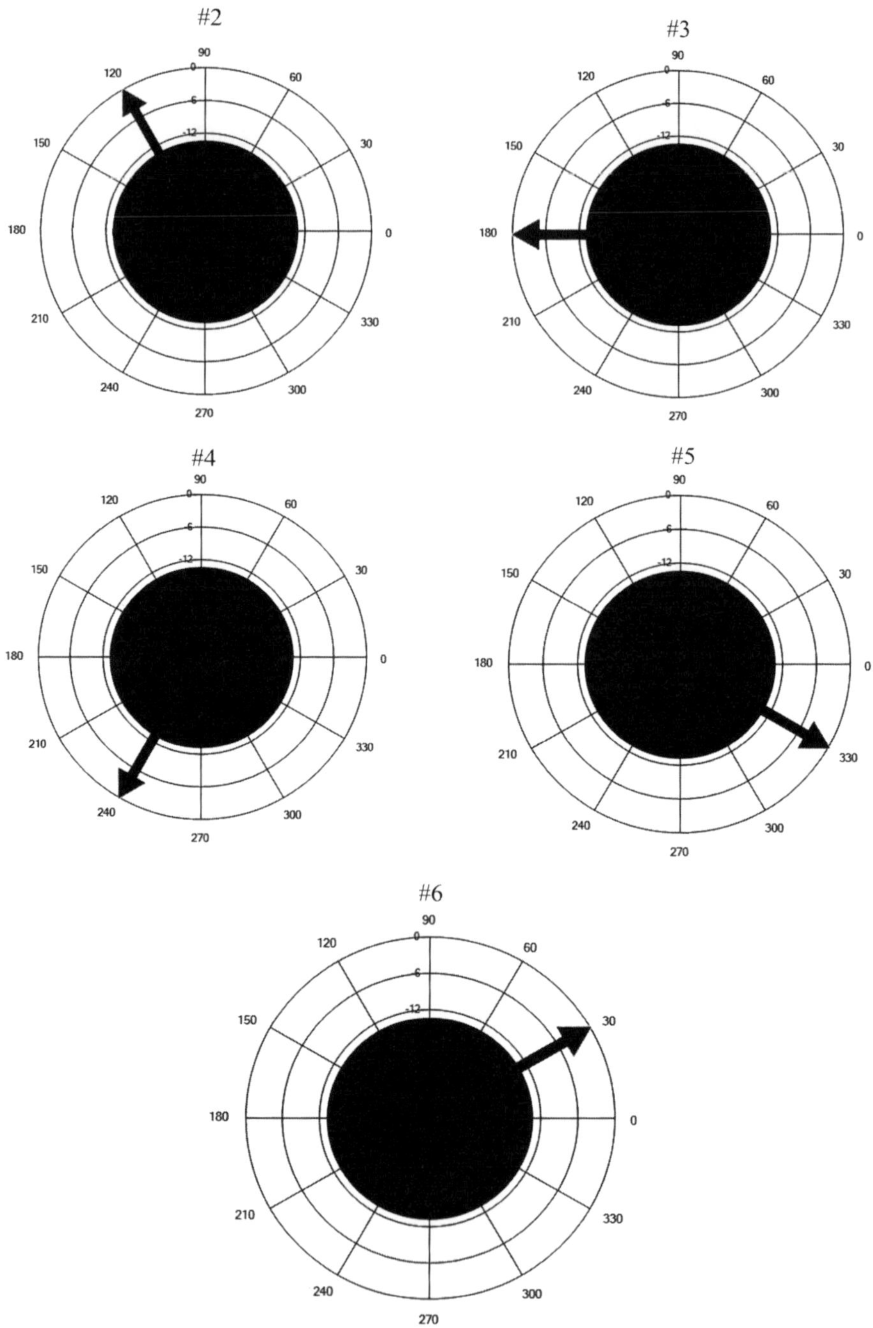

Fig. 5.13 Concept radiation patterns showing the angle radiation of different array elements

References

1. Chen Z, Luk K (2009) Antennas for base stations in wireless communications. McGraw-Hill, New York
2. Fujimoto K (2008) Mobile antenna system handbook, 3rd edn. Artech House, Norwood
3. Taeb A et al (2010) An overhead V-shape printed dipole array antenna for toll plaza application. Antenna and propagation international symposium, IEEE
4. Pan R et al (2008) A broadband printed dipole and a printed array for base station applications. Antennas and propagation international symposium, IEEE
5. He Q-Q, Wang B-Z, He J (2008) Wideband and dual-band design of a printed dipole antenna. IEEE Antennas Wirel Propag Lett 7:1–4
6. Basic Requirements for Global ETC., Ver. 1.0, July 2007 Global ETC Promotion Group. http://www.hido.or.jp/itsos/images/Basic%20Requirements%20for%20Global%20ETC%20Ver.%201.0.pdf
7. Luo B, Li P (2009) Budget analysis of RF transceiver used in 5.8 GHz RFID reader based on the ETC-DSRC national specifications of China. Wireless communications, networking and mobile computing, 2009, 5th international conference
8. Sherif MH (eds) (2009) Handbook of Enterprise Integration. CRC press, Boca Raton
9. Yoon J et al Fabrication and measurements of a microstrip array antenna for the electronic toll collection system (ETCS), Inha University, Korea. http://mnet.skku.ac.kr/data/2002data/CIC2002/cic2002/full_paper/paper_files/Paper_60.pdf
10. Bancroft R, Bateman B Design of a planar omnidirectional antenna for wireless applications, centurion wireless technologies, Westminster, Colorado. http://ram.lan23.ru/wifi/PCBCol/Collin_Microstrip_Planar.pdf
11. SuperPass Antenna Design Company. http://www.superpass.com
12. Huang J, Liang J (2008) Printed and double—sided dipole array antennas with a parallel reflector. Microw Opt Technol Lett 50(3):595–600
13. Donelly M et al (2007) A planar electronically reconfigurable Wi-Fi band antenna based on a parasitic microstrip structure. IEEE Wirel Propag Lett 6:623–626
14. Chuang H et al (1998) Omnidirectional horizontally polarized alford loop strip antenna. US Patent # 5767809, Publication Date 1998
15. Mak A, Rowell C, Murch R (2009) Low cost reconfigurable Landstorfer planar antenna array. IEEE Trans Antennas Propag 57(10):3051–3060
16. Pham N et al Microstrip antenna array with beamforming network for WLAN applications. http://www.ece.uci.edu/rfmems/publications/papers/antenna/C104-APS05.pdf, Department of Electrical Engineering and Computer Science University of California
17. Pham N et al (2005) Microstrip antenna array with beamforming network for WLAN applications. Antennas and propagation international symposium, IEEE, pp 267–270
18. Mahler W, Landstorfer F (2005) Design and optimization of an antenna array for WIMAX base stations. IEEE, international conference on wireless communications and applied computational electromagnetics, 2005, pp 1006–1009
19. Goshi D et al A single RF channel smart antenna receiver array with digital beamforming. http://www.ee.ucla.edu/~mwlab/poster/2004/Darren_ARR2004.pdf, University of California, Electrical Engineering Department
20. Goshi D et al (2004) A compact digital beamforming SMILE array for mobile communications. IEEE Trans Microw Theory Tech 52(12):2732–2738
21. Frederick J et al (2004) Smart antennas based on spatial multiplexing of local elements (SMILE) for mutual coupling reduction. IEEE Trans Antennas Propag 52(1):106–114
22. Frederick J et al (2002) A smart antenna receiver array using a single RF channel and digital beamforming. IEEE Trans Microw Theory Tech 50(12):3052–3058
23. Christ R (2009) Application and performance of personnel tracking systems. 43rd annual 2009 international carnahan conference on security technology, pp 120–128

24. Bullock JH (1998) Vehicle locator system. US Patent, 5786758, 1998
25. Aloi D, Sharawi M (2011) Automotive direction finding system based on received power levels. US Patent application publication #2011/0148578, Publication Date June 2011
26. Aloi D, Sharawi M (2010) Comparative analysis of single-channel direction finding algorithms for automotive applications at 2400 MHz in a complex reflecting environment. Phys Commun 3:19–27
27. Vu V, Delai A (2006) Digital solution for inter-vehicle localization system by means of direction-of-arrival. International symposium on intelligent signal processing and communication systems (ISPACS2006), pp 875–878
28. Bruzzone R (2001) Radio signal direction finding. US Patent 6,271,791, 2001
29. Peavey D, Ogunfunmi T (1997) The single channel interferometer using a pseudo-doppler direction finding system. In: Proceeding of 1997 conference on acoustic, speech, and signal processing, vol 5 (1997), pp 4129–4132
30. Harter N et al (2005) Development of a novel single channel direction finding method. In: IEEE proceedings of the military communications conference, vol 5, pp 2720–2725, Oct 2005
31. Rabinovich V (2011) Direction finding system for automotive applications using small phased antenna array. Microw Opt Technol Lett 53(10):2441–2446
32. Tieman A et al (2009) Vehicle locator key fob with range and bearing measurement. US Patent Application Publication, 2009/0264082, 2009
33. Rabinovich V, Rabinovich D (2010) Three port compact multifunction printed antenna system, antenna for automotive applications. Antennas propagation society international symposium IEEE
34. Wireless InSite User's Manual Version 2.3, Remcom, Inc, 2006
35. Cidronali A et al 2.45 GHz smart antenna for location-aware single-anchor indoor applications. Department of electronics and telecommunications, University of Florence, Florence, Italy. http://www.giannigiorgetti.com/media/PDF/ac_ims09.pdf
36. Cidronali A et al (2009) A 2.45 GHz smart antenna for location-aware single-anchor indoor applications. IEEE MTT-S international microwave symposium digest, MTT '09, pp 1553–1556
37. Giorgetti G et al (2009) Single-anchor indoor localization using a switched-beam antenna. IEEE Commun Lett 13(1):58–60

Chapter 6
Compact Car-Mounted Arrays

6.1 Design Examples for 2.45 GHz

6.1.1 Reactively Steered Ring Array Printed on Car Glass

As stated earlier in Chap. 1, relatively simple, reactively steered arrays include a main active antenna and a number of parasitic elements with electronically controllable reactance circuits. Figure 6.1a to c presents an example of 2.4 GHz reactively steered ring antenna array mounted on the vehicle glass [1]. The design does not degrade the appearance of the vehicle and saves installation space for other devices. Three overlapping ring-shaped wires, each one wavelength, are provided without any mutually nodal points. A feed point and five reactance circuits are embedded in the wires. By changing the values of the reactance circuit, beam direction can be controlled in xy plane. Reactance circuits are made of varactor diodes, the values of which are controlled by changing their input voltages. As described in [2], a one-wavelength ring antenna can be represented by two in-phase dipole antennas with a spacing of 0.27. Therefore, the ring element array is assumed to have an equivalent model as shown in Fig. 6.1d, which is composed of six dipole elements. Math simulation optimizing the antenna geometry for 2.4 GHz shows the following dimensions: $H = 39.8$; $D = 0.125$; $G = 312.5$; $L1 = 1.5$, and $L2 = 0.03$ (Fig. 6.1a). Figure 6.2 shows radiation patterns as a function of ϕ ($\theta = 90°$) calculated for beam direction toward ϕ=−90, −30, 0, 30, 90°. Calculations are performed using two different methods: MOM technique and equivalent dipole [2] presentation. Reference paper [1] also describes antenna prototype that is constructed on a typical Teflon substrate with thickness of 0.6 mm and dielectric constant of 2.6. Copper is etched on both sides of the substrate, and the reactance is realized using varactors, capacitors, and resistors.

V. Rabinovich and N. Alexandrov, *Antenna Arrays and Automotive Applications*,
DOI: 10.1007/978-1-4614-1074-4_6,

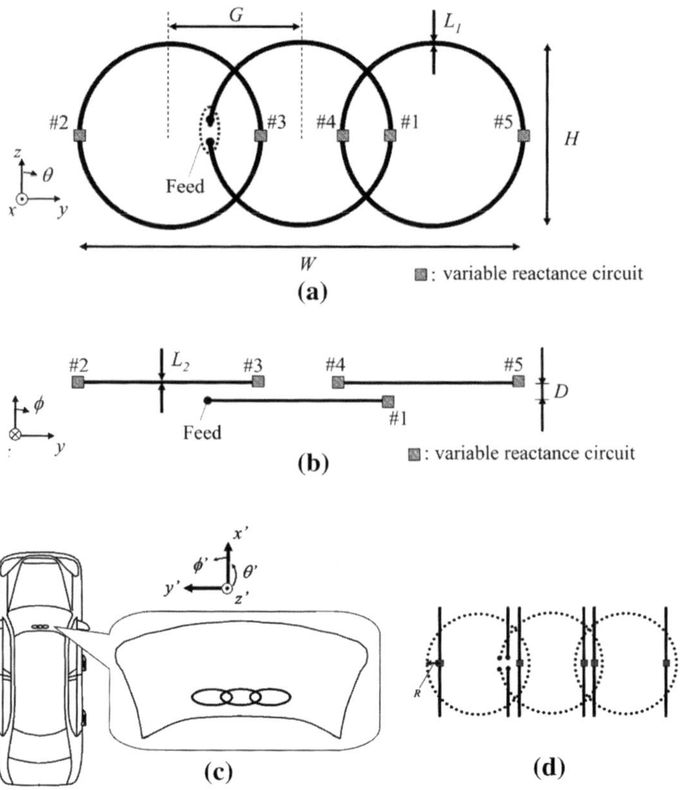

Fig. 6.1 Reactively steering array system printed on car glass. **a** and **b** Antenna array concept. **c** Array location on the glass. **d** Equivalent array representation, [1] © 2007 IEEE

The prototyped antenna is mounted at the upper center of the windshield, as shown in Fig. 6.1c. The distance between the metal roof top and the nearest side of the prototype is about 40 mm. The angle between the windshield and the horizontal plane is approximately 30°. Figure 6.3a and b shows the beam patterns toward the $\phi' = 10$ and $\phi' = 30°$ (coordinate system is shown in Fig. 6.1c).

Three patterns at every 10° of elevation angle from $\theta' = 90$ to $\theta' = 70°$ are plotted in Fig. 6.3a and b. Reactance values in ohm calculated by method of moment (MOM) are summarized at the Tables shown in Figs. 6.3c and d. It is seen that the radiation patterns are somewhat noisy due to the multiple reflections from the car body. The dimensions of the practically designed antenna module with bias circuits are $220 \times 50\ \text{mm}^2$. The antenna is composed of copper-etched multi-layered films. The feed ring wire and parasitic wire are placed on the independent films. The reactance circuits are mounted using conductive paste. The films are made of a transparent polyethylene (PET) substrate with the thickness of 125 μm.

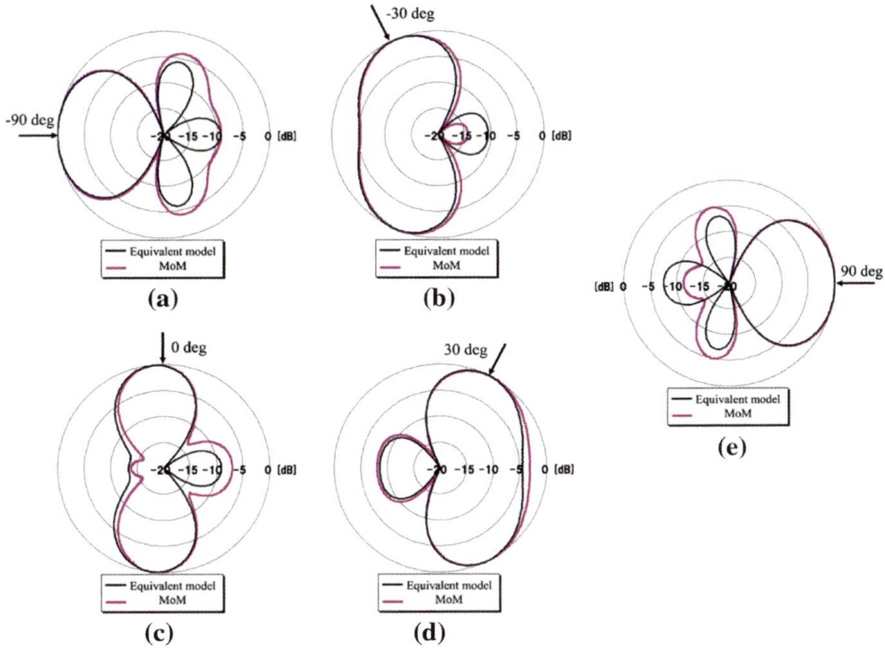

Fig. 6.2 Radiation patterns for different scan angles ϕ: **a** $\phi = -90$; **b** $\phi = -30$; **c** $\phi = 0$; **d** $\phi = 30$; **e** $\phi = 90°$, [1] © 2007 IEEE

6.1.2 Steerable Disk Patch Configuration

Detailed description of the disk patch antenna with parasitic elements is given in Ref. [3]. Figure 6.4a presents the schematic concept of the antenna system. The antenna occupies the area of less than 0.3 × 0.3 λ and is composed of a driven element (circular patch) and four surrounding parasitic elements. The central patch is driven by the inner conductor of a coaxial line. The outer conductor is connected to the ground plane. The bottom end of each T-shaped parasitic element is either short or open circuited to the ground plane (for simulation, the ground plane is chosen as an infinite). The main antenna dimensions for 2.45 GHz are as follows: the disk patch diameter is 29.4, patch and parasitic element heights above the ground are 9.8 mm, and horizontal length of each parasitic element is 24 mm. This compact array design, which could be easily mounted on the car roof, provides beam steering azimuth directions ($\phi = 0$, 45, 90, 135, 180, 225, and 315°). Steering of the angle in azimuth plane is realized with switches named in Fig. 6.4a as a, b, c, and d. It is found that the antenna array generates E_θ components from the vertical currents along the vertical central-driven pins, edge pins, and parasitic T element pins. Cross polarization component is <-30 dB.

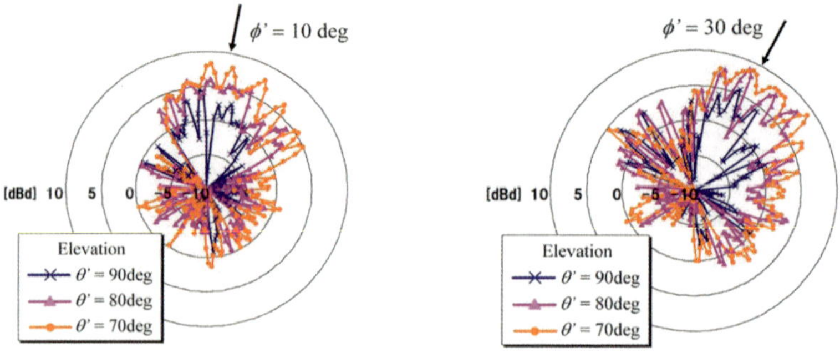

(a) **(b)**

X_1	-52
X_2	-2
X_3	-3.5
X_4	-56
X_5	-83

(c)

X_1	-4
X_2	-21
X_3	-37
X_4	-2
X_5	-1.6

(d)

Fig. 6.3 Measured radiation pattern for the array installed on the car: **a** Angle direction of the RF source $\phi' = 10$; **b** Angle direction of the RF source $\phi' = 30$; **c** Reactance values for $\phi' = 10$; **d** Reactance values for $\phi' = 30°$, [1] © 2007 IEEE

The antenna is a vertically polarized structure with a maximum E_θ value for θ=90° (horizontal plane). Figure 6.4 b to e shows the angles of maximum radiation as a function of the switch states a, b, c, and d. Any switch shown in the circle is the shorted switch and switch without the circle is in the open position.

6.1.3 Low Profile Vertically Polarized F-Antenna Array for Roof Mount Applications

Figure 6.5a presents *F*-type roof-mounted antenna array [4], which is a suitable candidate for automotive applications where compact design is required. The geometry of all elements is the same with the exception of one feature: each of the two actively excited identical elements (#1 and #2) has amplitude and phase control as shown in Fig. 6.5b and each of the two identical parasitic elements (#3 and #4) equipped with a varactor diode (Fig. 6.5c) can adjust the reactance value to provide beam steering. The antenna radiates vertically polarized waves

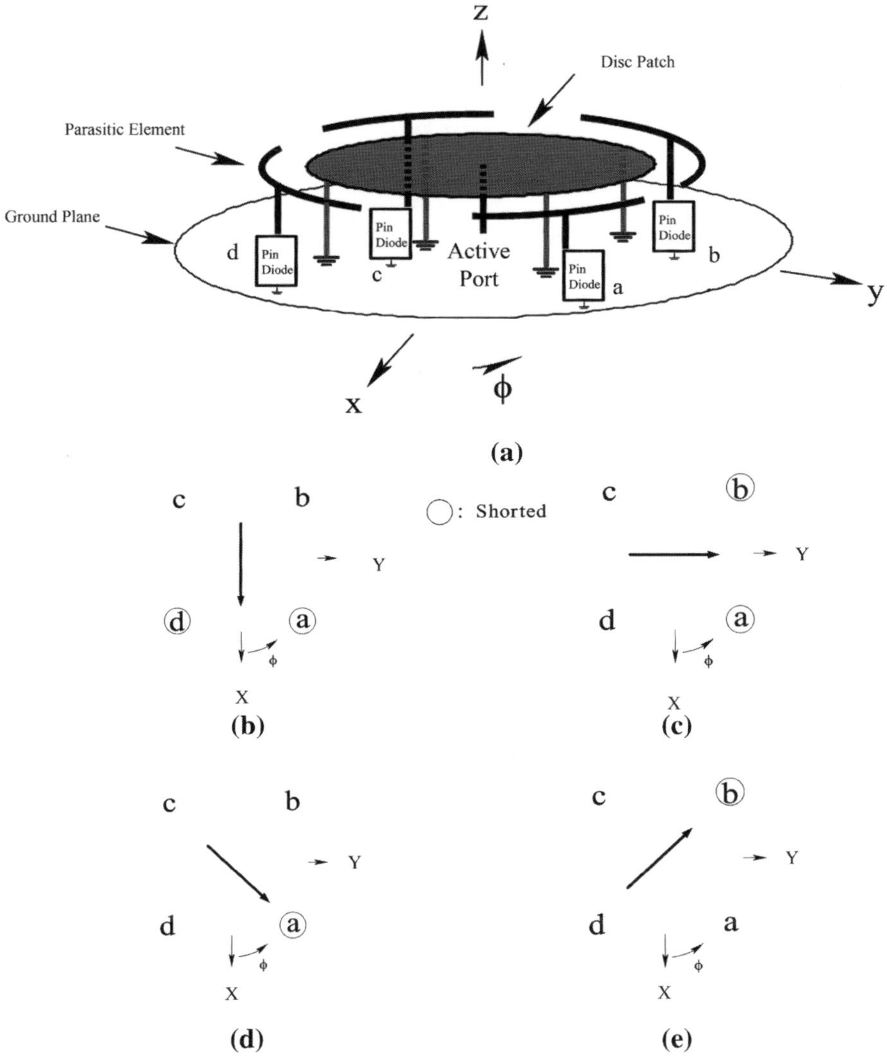

Fig. 6.4 **a** Schematic concept of the disk patch array configuration; **b** to **e** Angles of maximum radiation versus switch states

and can operate as an adaptive system, forming the desired angle direction of the main beam and nulls in the directions of the interference sources. The lengths of the individual components of the antenna element for 2.45 GHz applications are as follows: $h = 11$, $L1 = 17.8$, $L2 = 4.9$, and $Lp = 2.5$ mm. The lengths h, $L1$, $L2$, and Lp are adjustable, so that the imaginary part of the input impedance of the actively excited elements becomes 0 for a frequency of 2.45 GHz. Combined length $h + L1 + L2$ is approximately equal to $\lambda/4$. The total dimensions of the

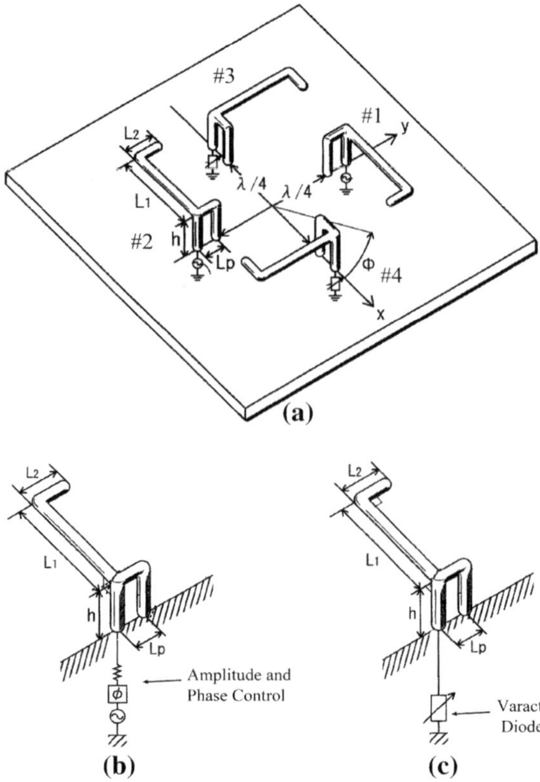

Fig. 6.5 **a** ESPAR F-type roof-mounted antenna array; **b** Active element with amplitude and phase control; **c** Passive element with varactor diode control

antenna are 35.5 × 35.5 × 11 mm^3. Figure 6.6 illustrates the calculated normalized, vertically polarized radiation patterns for the horizontal plane using numerical electromagnetic code (NEC). Referring to the figure, the unit of $A1$ and $A2$ is amplitude, ϕ_1 and ϕ_2 is degree, and X_3 and X_4 is expressed in ohm. As emphasized in reference source [4], antenna structure can form a beam toward arbitrary angle direction in the horizontal plane by properly adjusting the values of the excitation amplitudes $A1$ and $A2$, excitation phases ϕ_1 and ϕ_2, and reactance values X_3 and X_4.

When these values are adjusted, as shown in Fig. 6.7 the antenna system radiates beams and forms nulls in the directions marked in the figure by arrows. Figure 6.8 shows a few antenna array topologies described in the Ref. [4] which can also be beneficial for the car roof-mounted systems.

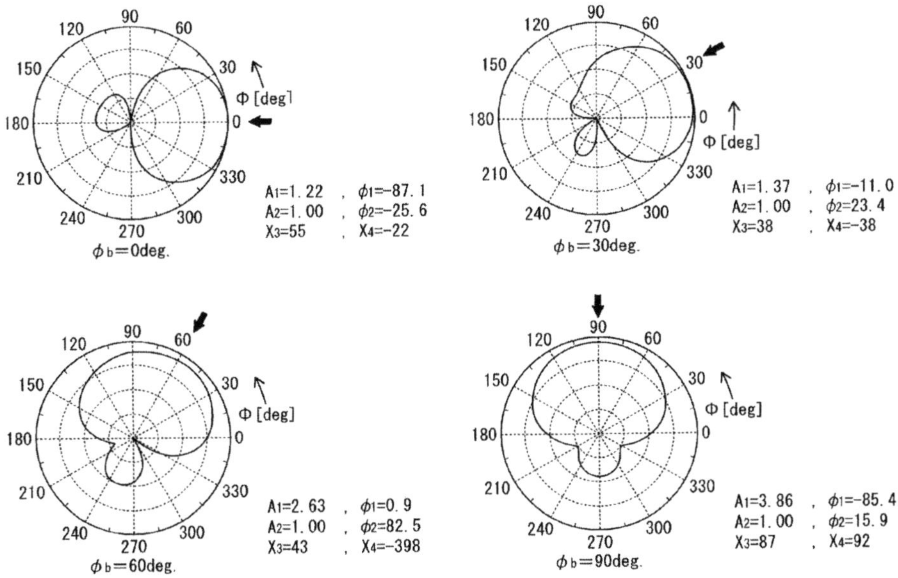

Fig. 6.6 Examples of scan radiation patterns for ESPAR *F-type roof-mounted antenna array*

6.2 Prototype Samples for 5.2 and 5.9 GHz Frequency Bands

6.2.1 Endfire Array for ITS Applications Providing Forward/Backward Radiation

A compact array configuration operating within frequencies from 5.85 to 5.925 GHz, schematically shown in Fig. 6.9a [5, 6], is another suitable candidate to generate directional beam radiation pattern for automotive applications.

The proposed array has scan capability in two directions, forward endfire or backward endfire. As shown in Fig. 6.9a, the array contains N number of L shaped wire monopole antenna elements mounted on the ground PC board. Bended elements are utilized instead of monopole antennas to decrease the overall height. The distance between the adjacent elements is around a quarter of wavelength, which is about 13 mm at 5.9 GHz. Each element has a length l of horizontal section and h mm of vertical section. The prototype of the antenna is based on the following parameters: number of antenna elements is equal to four and $h = 6$ mm. PCB is made of low-loss RF material with a dielectric constant of 2.3. Figure 6.9b presents the combination of two directional couplers to provide equal distribution of energy between four ports of the antenna array elements. They are numbered from #1 to #4 and two isolated feeding ports are numbered as Port 1 and Port 2. The feed network shown in Fig. 6.9b provides a 90° phase shift between adjacent elements. Therefore, when Port 1 is excited, energy radiates toward the element #4 and when Port 2 is excited, energy radiates toward the element #1. The array gain

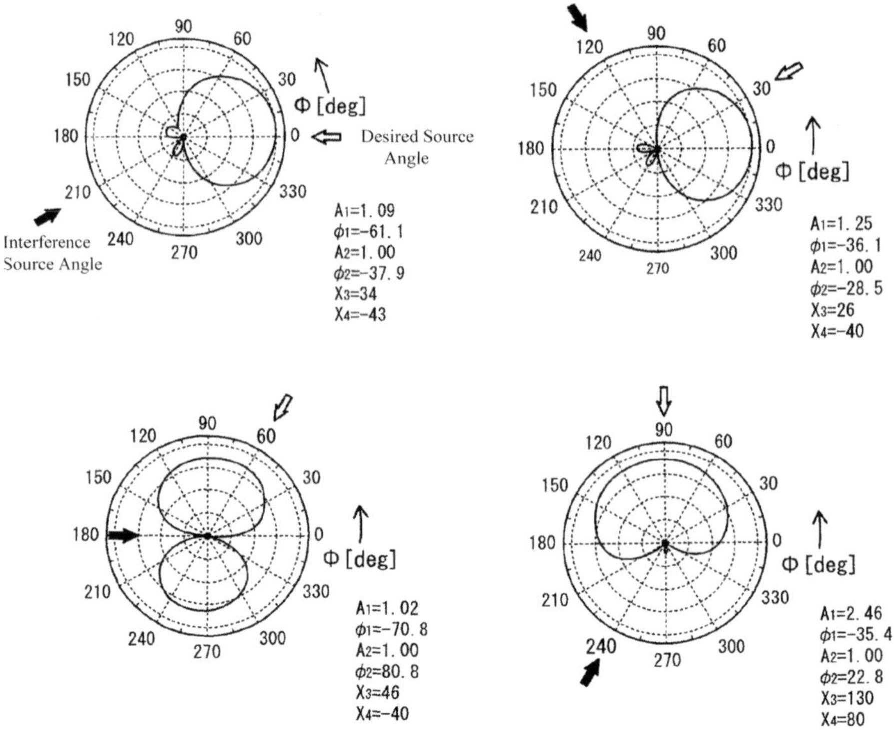

Fig. 6.7 Adaptive radiation pattern examples

is about 11.5 dBi and −3 dB beamwidth of the array is 82 and 40° in the azimuth and elevation planes, respectively. The table shown in Fig. 6.9 summarizes the antenna measurement parameters.

6.2.2 System Integrated into the Vehicle Body

The drawing of the patch antenna array integrated into the vehicle body is shown in Fig. 6.10a [7, 8]. Configuration presented in Fig. 6.10a imitates (roughly) an array inserted into the side car door and, therefore, allows radiating energy toward the angle directions #1 or #2 as shown in Fig. 6.10d. The prototype of the array described in the reference papers [7, 8] is composed of three radiating elements: the mid-patch is the transmitter antenna with port #1 and the other two are receiving elements with port #2 (Fig. 6.10c). The antenna element is built on the top side of the dielectric board while the bottom side is grounded.

Printed on FR-4 PCB with thickness 1.6 mm, the patch element has the following dimensions: length is equal to $L = 11.8$ and width $W = 15$ mm for 5.8 GHz, the inset feed distance from the patch edge is equal to 3.5, and the width

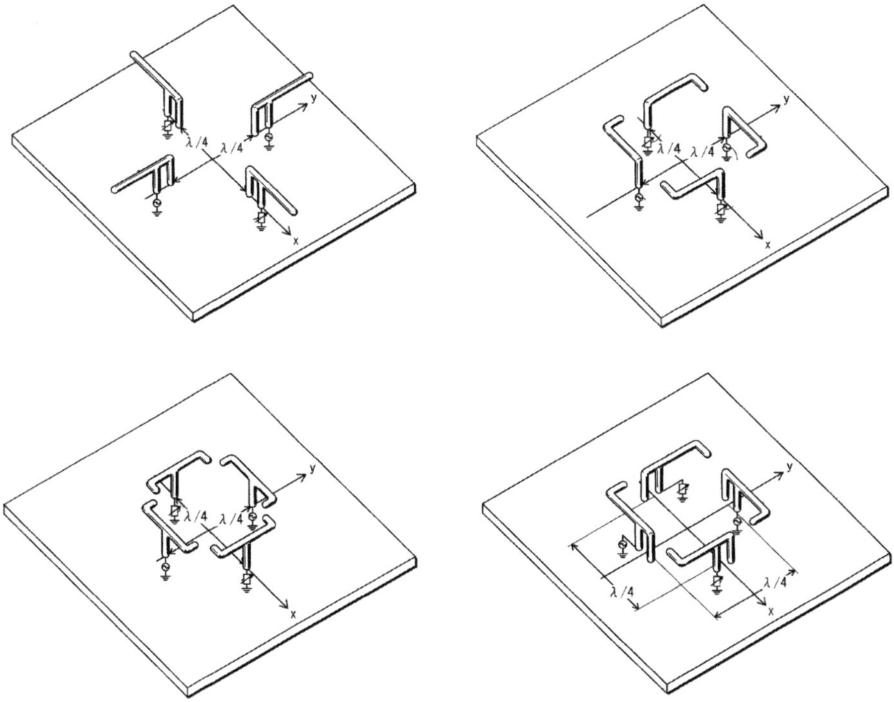

Fig. 6.8 Examples of ESPAR ray topologies

of the transmission line is 3 mm (Fig. 6.10b). The distance between adjacent elements is equal to 26 mm, which corresponds to the half of the wavelength in free space. It is shown that one mid-antenna element has the peak gain of about 5 dBi with main beam ripples of the order ± 2 dB and ripples of the sidelobes of about 5–7 dB due to the conducting plates. Two receiving antennas with spacing of one wavelength provide −3 dB beamwidth of about 60° and, the gain theoretically, two times higher than the gain of single patch element. The measured efficiency of the antenna in free space is equal to 0.75, the efficiency between conducting plates is about 0.72, and for the array in the dielectric case efficiency is equal to 0.62. These data are obtained for 5.8 GHz frequency. In a real environment, the car body shape generates ripples of the directionality, and peak gain level differs from the above noticed values. Therefore, final radiation pattern has to be measured experimentally for specific car model.

6.2.3 Compact Smart Topology for Wi-Max Radio

Figure 6.11a shows a compact, lightweight, uniform eight-element circular switched beam array antenna prototype [9]. The antenna system with a diameter of

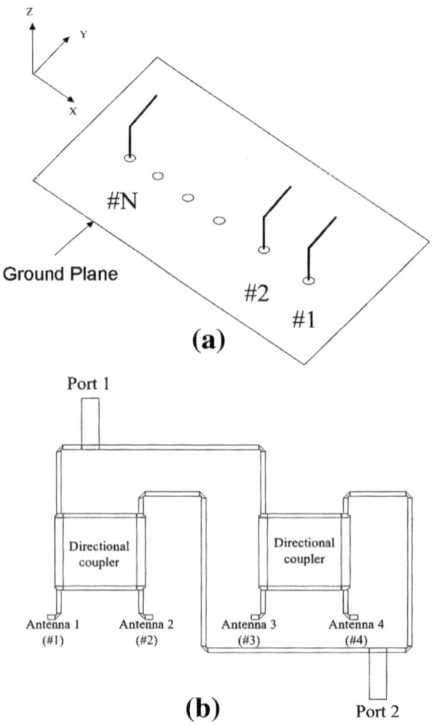

Frequency	5.85-5.925 GHz
Peak Gain	11.5 dBi
Return Loss	>15 dB
Polarization	Linear
3-dB Beamwidth, E_θ Horizontal Plane ($\theta = 90^0$)	82°
3-dB Beamwidth, E_θ Vertical Plane ($\phi = 0^0$)	40°
Total Dimensions of the PCB Ground (mm)	80×60
Element Material	Wire
Vertical Wire Portion (mm)	6
Horizontal Wire Portion (mm)	6
Number of Elements	4

Fig. 6.9 **a** Forward and backward endfire array topology; **b** Combining two directional couplers to provide equal distribution energy between four ports of the antenna array elements

75 mm covers a 360° field of view in the azimuth plane and operates at 5.8 GHz. In a switched beam regime array creates 16 fixed overlapping radiation patterns, as shown in Fig. 6.11b. The antenna unit contains two functional blocks: the circular monopole array with the radius of 24.6 mm and a beamformer. The main algorithm of the array has three main components: searching, beamforming, and tracking. The searching component collects the received signal from the connected radio at each predetermined angle position and selects the one desired signal with maximum received power. Once a desired signal from the known direction is chosen, the main beam of the antenna array can be tracked toward the variable arrival angle.

Let us assume that, beam #m is chosen. To update, we compare the signal strength from beam m and its neighbor beams: beam #$m-1$ and beam $m+1$, and choose the strongest one as the updated beam. The tracking cycle is chosen so that the target will not travel out of the small angle range between beam #$m-1$ and #m + 1. The reference paper [9] demonstrates the experimental results of the proposed array.

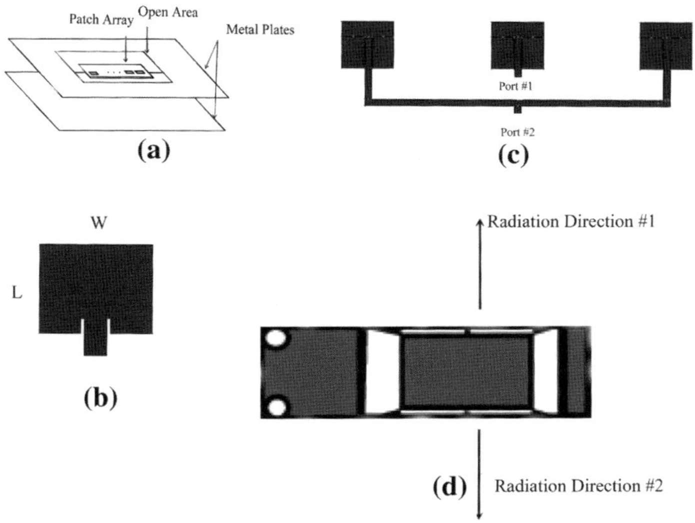

Fig. 6.10 **a** Concept of the patch antenna array integrated into the vehicle body; **b** Topology of individual array element; **c** Example of the patch array design: one transmitting and two receiving elements; **d** Radiation directions of the arrays inserted into the car side doors

6.2.4 Roof-Mounted Four-Element Patch Array for V2V Communication

An automotive antenna module consisting of four individual antenna elements for V2V/V2I communication [10] is demonstrated in Fig. 6.12a.

This system operates in the frequency band from 5.480 to 5.720 GHz, which is close to the main DSRC frequency 5.9 GHz [10, 11]. The antenna system can be utilized for MIMO applications with four individual output ports, as well as an antenna array with one combined output that realizes narrow scanning (if using additional phase shifters) beam or fixed beam antenna system. Due to the limited mounting space inside the module and with respect to a limited height, the antenna design is based on the low profile patch antenna elements. Four roof antenna elements form linear array separated by $\lambda/2$ at the operating frequency range. The Y-axis denotes vehicle driving direction. It is assumed that the antenna elements are integrated into the rooftop antenna module, which also includes six $\lambda/4$ parasitic passive monopoles, as shown in Fig. 6.12a. These elements are short circuited in their respective root points such that they act as discrete scatters in the antenna compartment. The total length of the antenna module L is 130 and width W is 50 mm. Individual antenna element sketch shown in Fig. 6.12b represents a short circuit ring patch operating at TM_{01} mode to provide a "monopole" radiation pattern [12]. The antenna prototype is manufactured using Rohacell dielectric material with a thickness of $t_{\text{patch}} = 1$ mm and dielectric constant of 1.0. The outer diameter of the circular patch is equal to F,

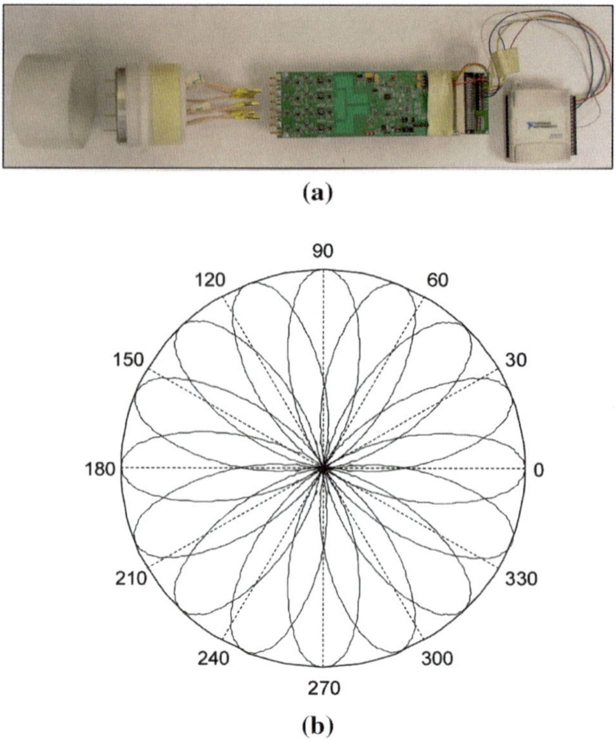

Fig. 6.11 **a** Compact adaptive circular monopole array for roof mount applications; **b** Multibeam radiation pattern, [9] © 2009 IEEE

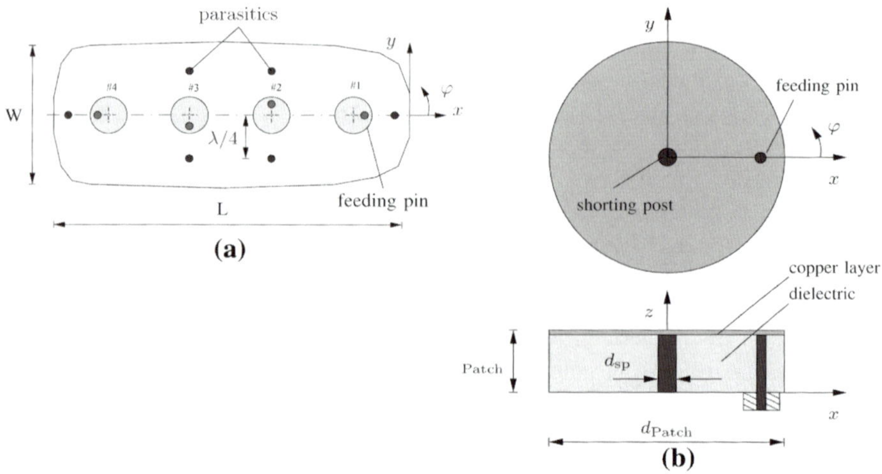

Fig. 6.12 **a** Antenna module with four individual antenna elements for V2V/V2I communication; **b** Individual antenna element topology

Table 6.1 Measured gain in dBi scale for different antenna elements (#1 through #4)

Antenna element #	G_1	G_2	G_3
#1	7.1	9.1	7.8
#2	7.5	9.6	6.8
#3	5.7	9.5	7.5
#4	4.8	7.8	3.4

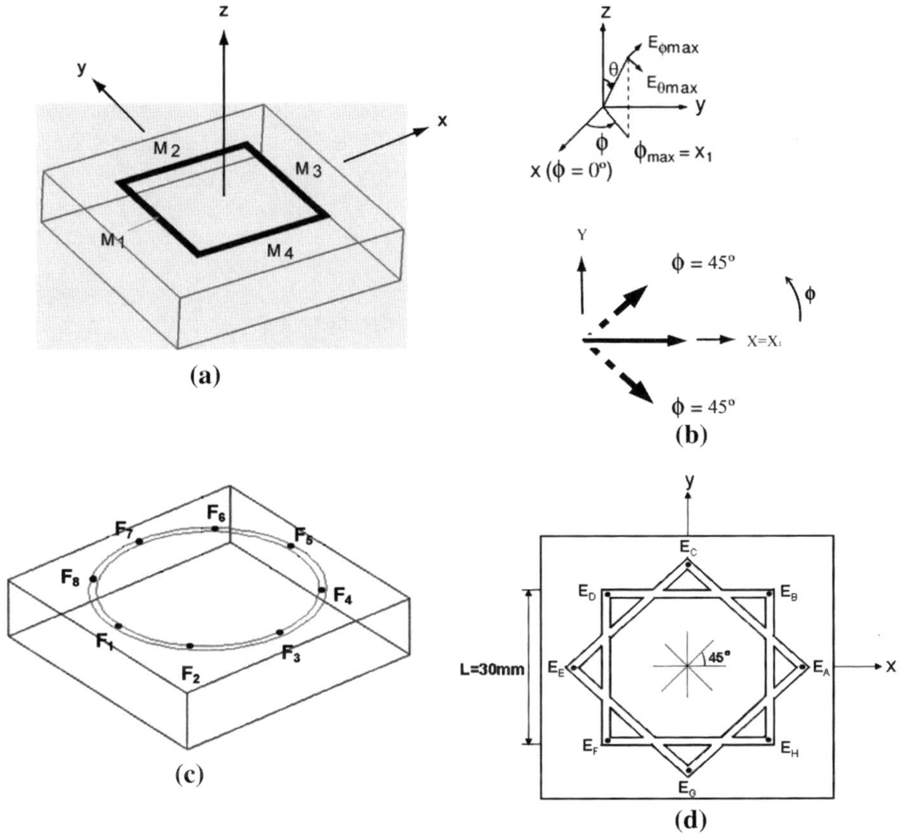

Fig. 6.13 **a** Square loop array design; **b** Angle directions of the radiation pattern maximum in azimuth plane, components E_θ, and E_ϕ are marked by solid and dashed arrows, respectively; **c** Circular array configuration; **d** Combined two square loop topology

d_{patch} = 10.0 mm, and the shorting post thickness d_{sp} = 1.0 mm. Scattering parameters, which estimate mutual coupling coefficients between all four individual antennas, are below −12 dB in the entire frequency band of operation. They are measured for the roof antenna module mounted on a small electric ground plane with the diameter of 200 mm. Impedance matching at 50 ohm provides return loss for all elements <−10 dB.

Table 6.1 demonstrates measured gain in dBi scale for different antenna elements (#1 through #4) under the following conditions:

The gain G_1 corresponds to the antenna module mounted on a sophisticated custom-made ground plane with a diameter equal to 1 m, rolled edges in order to reduce diffraction effects of the electromagnetic field, and without parasitic pins;

The gain G_2 is obtained when the antenna module includes shorting pins, and the ground plane has a 1 m diameter;

The gain G_3 is the gain estimation for the antenna module mounted on the car roof.

All results are demonstrated for the vertical component of the electric field, tilt angle θ=75°, and correspond to the maximum values obtained within azimuth angle 0–360°.

Due to the passive scatters gain of all elements are increased and as indicated in reference paper [10], gain distribution over 360° becomes more uniform. Finally, the realistic vehicular antenna integration and roof mounting significantly affects the gain of proposed solution.

6.2.5 Low Profile Beam Switched Loop Design

The schematic topology of a few compact loop antenna arrays [13, 14] with beam steerable capability is presented in Fig. 6.13a, c, and d.

The loop antennas demonstrated in Fig. 6.13 make use of a number of equally spaced feeds. Upon activating one feed at a time, the beam can be steered from one point to another, thus, realizing a steerable antenna. The loop antenna is etched on the printed circuit board with a ground plane on the bottom side and a metal strip line on the upper side. Each port is activated with pin-diodes providing beam switching over a 360° azimuth.

The square loop antenna shown in Fig. 6.13a feeds from ports $M1, M2, M3$, or $M4$ and can produce four fixed beams steering in 90° steps. The antenna built on the PCB with a dielectric constant of 3.45, dielectric thickness of 12, square side length 30, and loop conductor width 1.5 mm has two resonance frequencies: 3 and 5.2 GHz, which is popular for Europe WLAN communication. For example, an active port $M1$ generates a titled E_θ radiation pattern component along $\phi_{\max} = 0°$and $\theta_{\max} = 45°$ at 5.2 GHz. The beam also has a $E_\phi$component with two maximums for $\theta_{\max} = 45°$, ϕ=45, and ϕ=−45°. Directions of the maximum components E_θ(solid line) and E_ϕ (dashed line) for cut $\theta_{\max} = 45°$ in the ϕ plane are shown in Fig. 6.13b.

The circular loop design presented in Fig. 6.13c has eight ports $F1$, $F2$... $F8$ and provides eight fixed beams with 45 steps over a 360° azimuth using pin-diode switches. For the inner loop radius equal to 19.09, strip width equal to 1.5, $h = 12$ mm, and dielectric constant of PCB $\varepsilon_r = 3.45$, antenna resonates at the frequency 4.4 GHz. The radiation pattern for activated port $F1$ generates beam directed to the following angles: $\theta_{\max} = 45°$ and $\phi_{\max} = 0°$.

The switched beam antenna shown in Fig. 6.13d is combined with two squares. The antenna has the following parameters: $L = 30$, $w = 1.5$, $h = 12$ mm, and $\varepsilon_r = 3.45$. Antenna has two resonant frequencies: 2.7 and 4.3 GHz. The beam for activated port E_a is directed toward $\theta_{\text{max}} = 45°$ and $\phi_{\text{max}} = 180°$. When all ports are activated the radiation pattern has eight beams covering 360 in the azimuth plane with an angle shift equal to 45°.

6.2.6 Low Profile Omnidirectional Horizontally Polarized Topology

This section presents conical beam array (Fig. 6.14) suitable for high performance radio local network (HIPERLAN) between the vehicle and the base station. The radiation pattern is omnidirectional in the horizontal plane with horizontal polarization and directional in the elevation plane with zero value toward zenith angle direction [15]. Figure 6.14a and b shows two possible array configurations: three- and four-patch antenna elements arranged in a ring formation. Designed antennas are fabricated on PTFE substrate with a dielectric constant $\varepsilon_r = 2.55$, dielectric material thickness $h = 1.524$, and loss tangent of 0.0024. If it is assumed that the patches are equally excited with electric current components in the direction of increasing ϕ, there will be null radiation in zenith, and the far field components should be horizontally polarized waves. Each patch of the three-antenna system has output impedance 150, which suggests a simple approach of connecting 150 lines at the patch edge, without any inset. Such lines provide 50 ohm match at the antenna system output.

The four-patch antenna array uses 4.9 mm inset on the radiating edge to obtain 80 ohm resonant impedance, and 80 ohm microstrip lines are attached at these points. At the center pin, each line has a quarter wavelength transformer raising the impedance to 200 ohm, and the antenna system output again has 50 ohm output impedance. The sizes of the finite ground plane are selected.

A set of measured patterns for three-patch array (75 × 75 mm) is presented in Fig. 6.14c and d, the radiation patterns for four-patches (80 × 80 mm) are demonstrated in Fig. 6.14e and f. It is clearly seen that the measured radiation patterns have null toward zenith and almost omnidirectional patterns over a 360° in the horizontal plane (peak-to-peak variations are 2 dB).

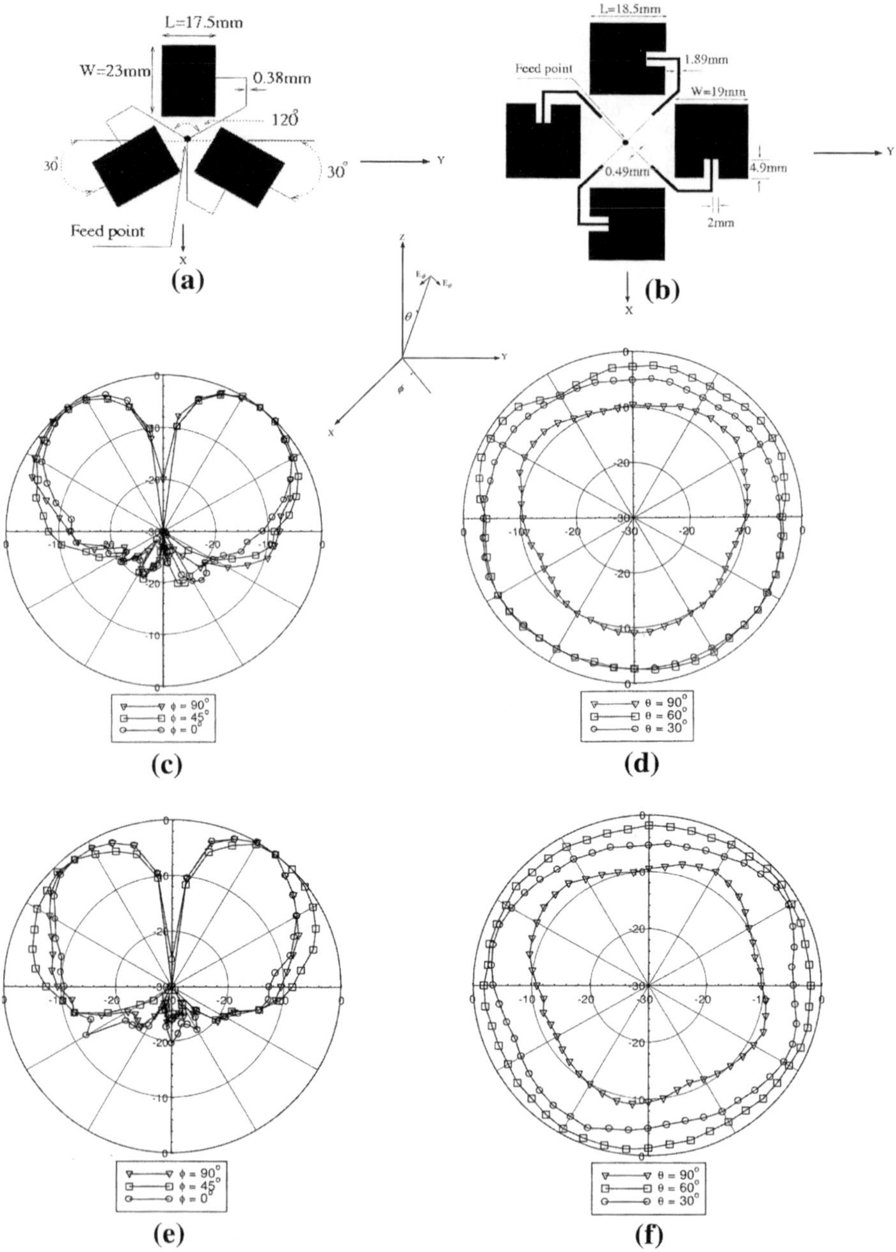

Fig. 6.14 **a** Three patch array arranged in ring formation; **b** Four patch array; **c** Elevation radiation pattern for three patch array; **d** Azimuth radiation pattern for three patch array; **e** Elevation radiation pattern for four patch array; **f** Azimuth radiation pattern for four patch array, © 2003 IEEE

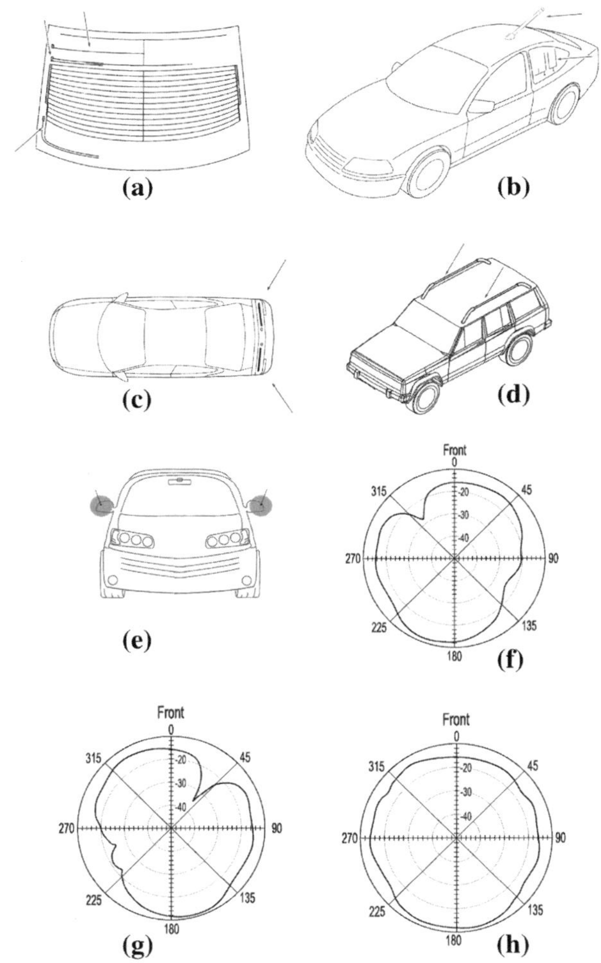

Fig. 6.15 Space diversity systems for car applications. **a** Rear side window glass array; **b** Side glass array concept; **c** Rear spoiler design; **d** Roof rails configuration; **e** Mirror integrated system; **f** and **g** radiation patterns for single antenna element; **h** Diversity radiation pattern

6.3 Broadcasting Vehicle-Mounted Multi-Element Systems

6.3.1 FM Radio Space Diversity Topology

Figure 6.15 shows (by the arrows) potential FM diversity antenna locations in the interior or exterior parts of the car. Figure 6.15a demonstrates three-element array printed on the car glass. Usually, such antennas exhibit gain in the range of 7–15 dB less than the gain of the regular whip system and, therefore require low noise amplifier application. The typical gain of the amplifier varies from 6 to 15 dB, depending on the passive antenna portion gain. Noise figure of the amplifier is about of 2–4 dB. The antennas can be mounted on the rear or side glasses, as shown in Fig. 6.15a and b.

When antennas occupy small glass area and have small output impedance value (much <50 ohm) in the entire FM frequency range, it is possible to utilize a high input impedance amplifier, which provides low output noise of the system [16, 17]. Figure 6.15b shows the option of the diversity system, which uses the combination of the regular roof antenna design with a printed on the glass solution. Figure 6.15c demonstrates two hidden diversity antennas mounted on the car rear spoiler [18]. The system mounted in the roof rails equipped with the phase space diversity system, is described in the reference patent [19] and shown in Fig. 6.15d. Design provides the combined diversity technique, that is presented in Chaps. 1 and 4. The antenna elements are separated by the distance of about half of the wavelength, and, therefore, multipath fading signals impinging to the antenna elements are practically uncorrelated. Figure 6.15e shows in mirror integrated diversity antenna elements. This compact solution may be applied for the FM diversity system [20], as well as for SDARS, Cell phone, or TV diversity design. Figure 6.15f, g, and h illustrates the conceptual radiation patterns for the two antenna elements diversity system: the first figure is related to the first antenna element, the second figure shows the radiation pattern of the second antenna, and final third figure is a combined (due to operation of the diversity system) radiation pattern. As we can see, the combined radiation pattern is almost omnidirectional while pattern measured for each antenna has dips more than 15 dB. A designer who starts to build an antenna diversity system must achieve a compromise between price and quality of the reception. For example, requirements for TV diversity reception system have to be more stringent in comparison with FM radio because it receives voice and video signals.

6.3.2 Examples of Diversity Systems for Digital Terrestrial TV

Digital video broadcasting (DVB) is a new combination of various digital technologies that provides better quality pictures in comparison with an analog TV. More and more countries are using DVB to transmit television and broadcast programs. As we mentioned before, DVB covers the VHF band from 110 to 270 MHz and the UHF band from 350 to 870 MHz (depending on the country). Broadband, low cost, and a compact antenna system are key for high-quality TV reception in a vehicle. Generally, mobile TV antennas have some problems when receiving TV signals, compared to the household fixed antenna systems. A TV mobile antenna is installed lower in comparison to a typical household fixed antenna, and as a result, the received signal is lower. It is known [21] that the correction factor for received electric field strength in the UHF band for reception at 10 and 1.5 m above the ground is 16 dB. This means that the received electric strength for the lower mounted antenna could be much less in comparison with the high-mounted antenna design. In addition, the signal received by a moving antenna suffers from significant fluctuations. Figure 6.16a shows an example of variation of the 509-MHz frequency signal (the bandwidth is about 430 kHz) received while

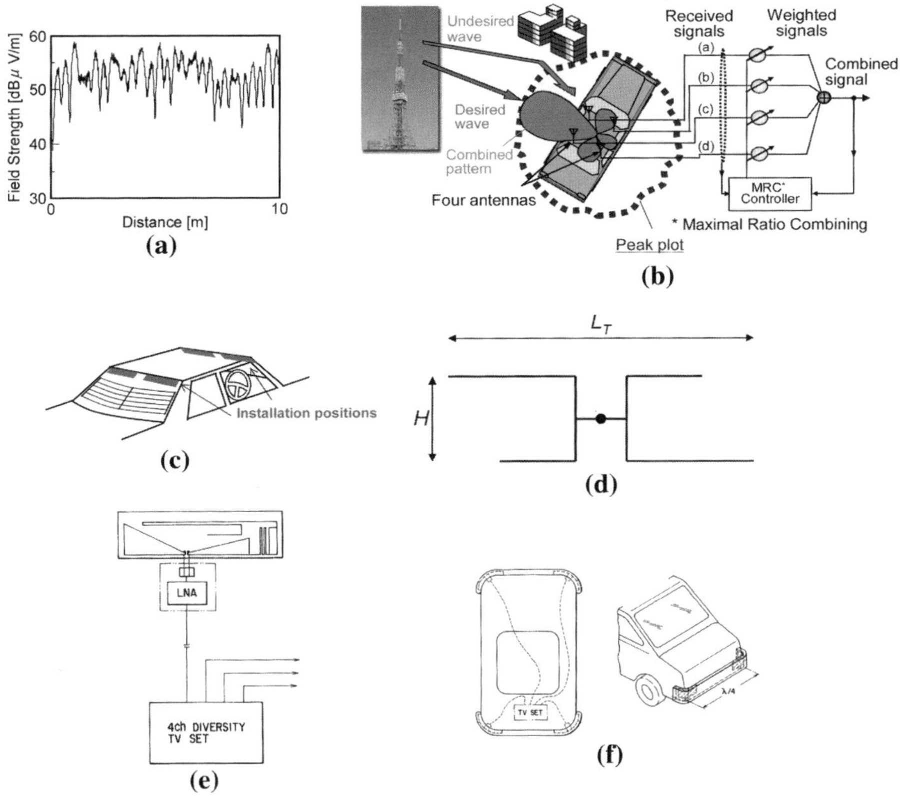

Fig. 6.16 Examples of diversity systems for terrestrial TV. **a** Fluctuations of the signal recorded by the receiver installed in the moving car; **b** Communication scenario between base station and the car in the conditions of multipath propagation; **c** TV array mounted on the front and rear car glasses; **d** Car glass antenna array element; **e** Array element printed on flexible foil; **f** TV diversity system mounted on the car bumpers

moving at 36 km/h. The figure shows that the electrical field strength drops many times during the short traveling distance of 10 m. Therefore, the typical antenna system for TV terrestrial applications uses the diversity technique, and each antenna element has a low-noise amplifier to increase the dynamic range of the receiving signals. A broadband amplifier has to provide signal amplification in a wide frequency band with minimal inserted noise. A high input impedance amplifier [22], which has a gain of about 15–20 dB in the entire frequency band, can be a suitable candidate for broadband TV applications.

Figure 6.16b shows a communication scenario between the TV base station and the car in the multipath environment. As noticed in Chap. 3, the best diversity solution in such a situation is the maximum ratio combining technique, which is used to design the antenna system consisting of four antenna elements printed on the front and rear car glasses, as shown in Fig. 6.16c [23]. The dimensions of the

Table 6.2 (I) Array parameters for Direct Broadcast Satellite (DBS) TV antenna systems

Parameters	Europe requirements (NATALIA Project)	US requirements
Frequency	10.7–12.75 GHz	12.2–12.7 GHz
Polarization	Linear with electrical control	Circular: LHCP/RHCP
Coverage angles	360° in Azimuth, 20–70° in Elevation	360° in Azimuth, 15–65° in Elevation
Peak gain(without LNA)	>20 dBi	>30 dBic
3-dB Beamwidth in both angle directions	~ 10.3°	~2.7°
Cross polarization isolation	>15 dB	>25 dB
Dimensions	200 Diameter, 20 mm Height	600 Diameter, 50 mm Height
Gain of low noise two stages amplifier (LNA)	12–15 dB (including phase shifter loss)	12–15 dB (including phase shifter loss)
Noise figure	1–1.5 dB	1–1.5 dB
The ratio of the gain to the noise temperature (G/T-A figure of merit)	>–6 dB/K	>10 dB/K

single antenna element shown in Fig. 6.16d are $L_T = 257$ and $H = 60$ mm. The system prototype based on the design with four antennas and combined diversity technique [23] generates almost omnidirectional radiation pattern with variations of about 3 dB over 360°.

The diversity system configuration printed on the flexible foil [24] is shown in Fig. 6.16e. Four identical antennas are mounted on the front and rear bumpers of the vehicle, as schematically demonstrated in Fig. 6.16f. The antenna element dimensions are 470 by 100 mm. The typical folded dipole intended for TV reception in the VHF band (90–222 MHz in Japan) is about 1,300 mm long. The suggested antenna is 830 mm shorter than the folded dipole. The designed antenna has a balun that transforms the input impedance of 300 ohm to the output impedance equal to 50 ohm. The measured VSWR in the 90–108 MHz band is <4:1, in the 170–220 MHz band <6:1, and in 470–770 MHz band <4:1. As shown from experimental results, the minimal antenna sensitivity for a horizontally polarized wave for any of the four bumper-mounted antenna elements is more than 40 dBu. Each antenna element has a low-noise amplifier, which provides a high-quality diversity system.

6.3.3 Satellite TV Arrays

Generally, a satellite TV antenna is mounted on the roof of the vehicle. The satellite receiving system tracks the signal received by an antenna from the satellite transmitter when the car is in motion. Therefore, the system has to be equipped with a special sensor to know where to find the satellite with minimal

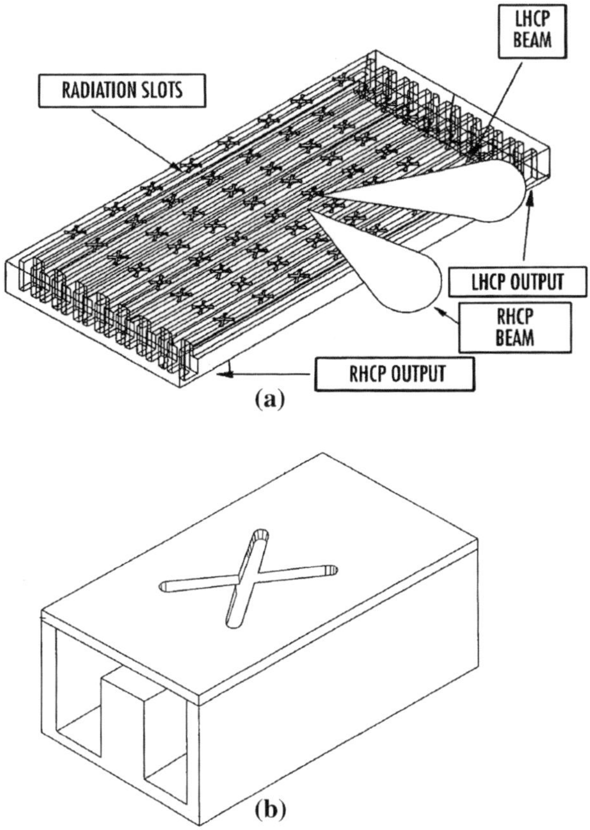

Fig. 6.17 **a** Ridged waveguide antenna array concept; **b** *X*-shaped cross-slot radiating element

searching. The tracking process has to keep the maximum radiation antenna pattern at the angle position corresponding to the satellite angle direction [25–28] and it has to meet certain requirements. Table 6.2 summarizes the main array parameters for Direct Broadcast Satellite (DBS) TV antenna systems intended for European and North American applications [25, 31].

European new tracking antenna project (NATALIA) [31] investigates the compact cost-effective receiving full electronically steerable antenna array for automotive applications. The system has to communicate with broadcast satellites emitting dual orthogonal linearly polarized signals. For a mobile receiver, these polarizations may appear at arbitrary orientations. In accordance with this condition, the array has to be configured to reorient the polarization electronically. For North American applications [25], the polarization of the signal is switchable between right-hand circular polarization (RHCP) and left-hand circular polarization (LHCP). An array for North America has to provide the following tracking parameters: elevation tracking rate of 20°/s with an acceleration of $50°/s^2$ and in the azimuth, the tracking rate must be 45°/s with an acceleration rate of $60°/s^2$.

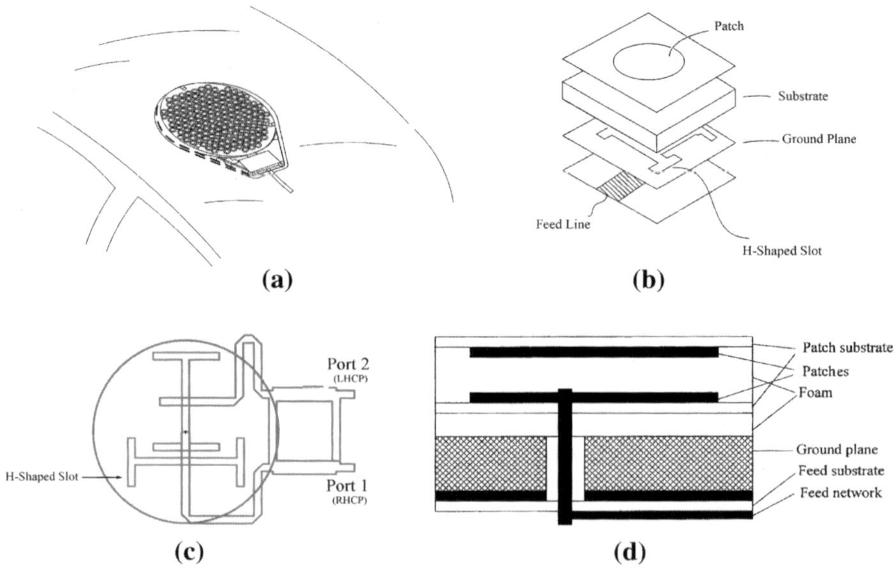

Fig. 6.18 **a** Common size of the NATALIA array; **b** Antenna element topology; **c** Feeding network for single element; **d** Side view of the radiating element with increased bandwidth

6.3.3.1 Ridged Waveguide Antenna Array with Circular Polarization

An antenna design using a ridged waveguide [26–28] with an electronic scan capability in the elevation plane is shown in Fig. 6.17a. The array consists of 32 ridge waveguides, with RHCP and LHCP outputs located on opposite ends of the antenna. A single waveguide with a radiating element is shown in Fig. 6.17b. The waveguide couples the energy from all radiating elements in the waveguide and combines the energy together. The radiating surface of the waveguide antenna uses multiple radiating elements distributed at uniform spacing along the waveguide axis. In the presented antenna, the radiating elements are *X*-shaped cross-slots, as shown in Fig. 6.17b.

The phase centers of the cross-slots are positioned along a straight line along the waveguide axis and between the center line of the waveguide ridge section and one of the walls. The radiation elements are placed about half a waveguide wavelength apart. The width of the waveguide is about 2 cm, the height is about 1 cm, the ridge height is around 0.7 cm, and the length of the radiating element is about 1 cm. The total length of the waveguide is about 60 cm with around 30 cross-slot elements spaced 2 cm apart. The angle beamwidth of one waveguide at the plane of the waveguide location is around 2.3–2.4°. RHCP and LHCP outputs are shown in Fig. 6.17a. Each of the 32 waveguides has an antenna output "probe" located on each end of the waveguide. The antenna "probe" is used to couple the output signal from the waveguide to the beamforming combiner, which adds

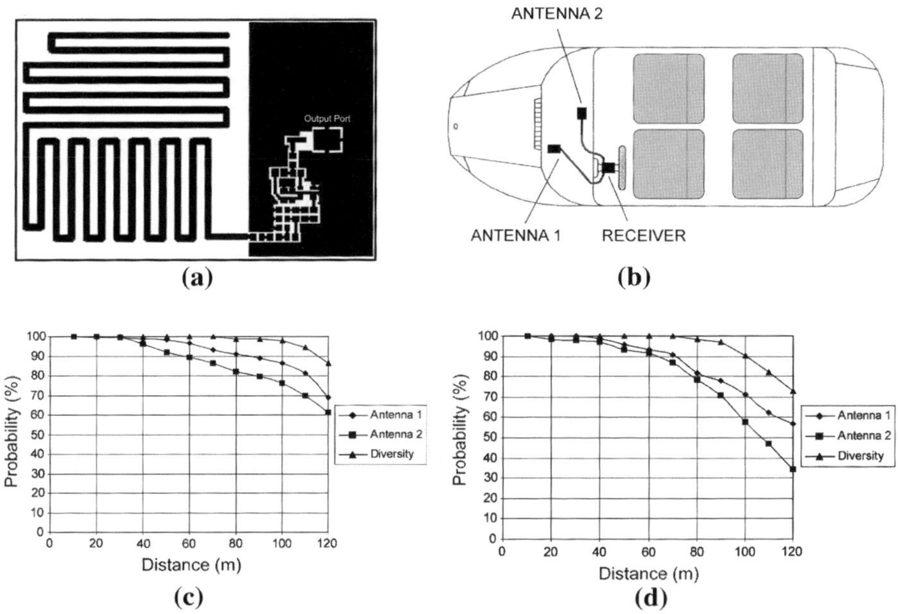

Fig. 6.19 **a** Meander line printed antenna element; **b** Two array elements diversity system concept; **c** Probability curve, based on the measurement results, vertical polarization; **d** Probability curve, based on the measurement results, horizontal polarization

signals from all 32 waveguides, providing the maximum radiating energy toward the TV satellite. Each antenna "probe" has a low-noise amplifier, and (depending on the structure of the beamforming network structure) a passive combiner with the output connected with a single receiver or an active combiner with a receiver in each antenna "probe". The antenna design has to have a small profile (<60 cm) and to be embedded in the vehicle roof, or between the roof and the headliner, while meeting the requirements for DBS antennas.

6.3.3.2 Linear Polarized Patch Antenna Array: NATALIA Project

The planar array shown in Fig. 6.18a is composed of 156 circular polarized elements arranged in a hexagonal grid. Figure 6.18b presents the simplified topology of the element realizing linear wave polarization [29–31]. With reference to Fig. 6.18b antenna array element consist of a circular patch printed on the substrate, ground plane with H-shaped slot line, and the microstrip feed line, which provides patch excitation.

The ground plane, located between the microstrip line and the radiating element, is configured to communicate signals through the slots only, separating the feed circuit layer from the radiating circuit layer. Complete decoupling of the feed

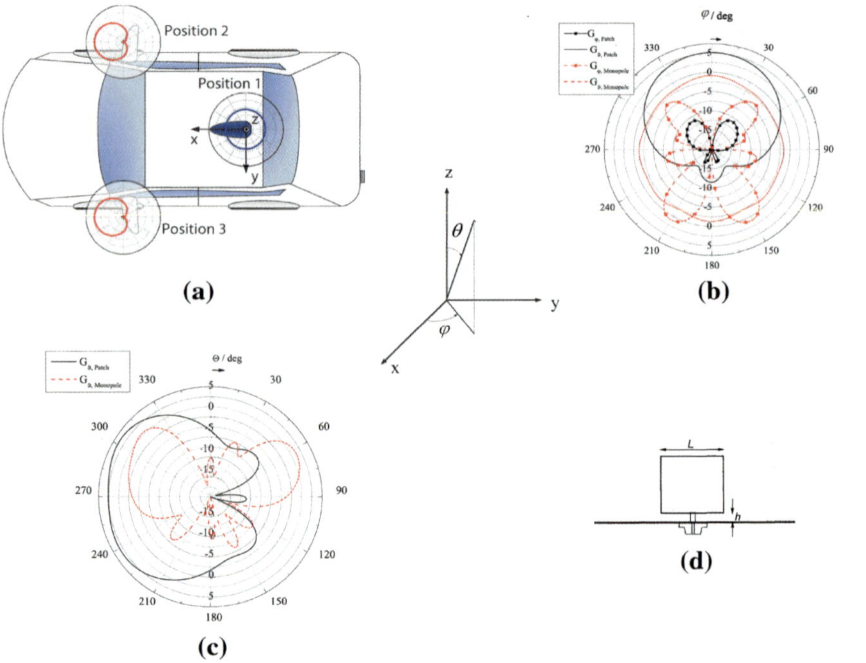

Fig. 6.20 **a** Roof-mounted distributed vehicle array for MIMO applications, two patch antennas, and one monopole position on the roof center; **b** Gain pattern of patch element in azimuth plane (dBi), $\theta = 90°$; **c** Gain of patch element in elevation plane (dBi), $\theta = 0°$; **d** Single element dimensions, [33, 34] © 2010 IEEE

line layer from the patch avoids unwanted radiation of the feeding and, therefore, improves the antenna parameters. The benefits of using *H*-shaped slot includes more compact size compared to the linear slot and offers a smaller required surface for coupling with the patch. Figure 6.18c shows the concept of the feed network that provides two orthogonal linear polarizations. A 90° hybrid coupler converts them into two circularly polarized waves. Figure 6.18d demonstrates the side view of the radiating element with increased bandwidth of the antenna element [29]. This configuration with two stacked patches positioned at the certain spacing allows for the increase in bandwidth from 5 (single patch) to 20 %, according to the system requirements. LHCP and RHCP wave components are applied to the microwave monolithic integrated circuit (MMIC) with low-noise amplifiers, digital phase shifters, and polarization combiner of two circularly polarized wave components into a linear polarized wave with electronically adjusted tuned angle direction. The antenna array uses 4-bit digital phase shifters, which realize phases with the discrete equal to 22.5°.

6.4 Diversity Design for RKE/RSE/TPMS

As stated in the earlier sections, the communication range that can be achieved in a radio system depends strongly on the antenna gain. As previously discussed, hidden antennas invariably suffer from performance problems due to shielding from the vehicle. Extended range for hidden automotive antennas can be a problem for some angle directions. For some designs maximum to minimum antenna directionality values over 360° show differences of up to 30 dB. This ratio is equivalent to about ten times the range ratio. Let us assume that the maximum radiation pattern value of the RKE system corresponds to the range 100, while the minimum directionality value corresponds to the range <10 m. In this case, one of the most effective ways to increase the range is to use the antenna diversity technique. A highly efficient, small size, diversity antenna system [32], integrated with a low-noise amplifier and based on the meander line antenna operating at 315 MHz, is presented in Fig. 6.19.

A 315 MHz band meander line antenna with linear dimensions 70 by 50 mm is shown in Fig. 6.19a. Figure 6.19b presents a diversity system with two RKE/RSE/TPS antennas placed under the front dash panel. Generally, the quality of the keyless entry system can be estimated using the probability $P(D > L)$that the range D is more than a certain value L. This probability is the ratio of the number M of the points for which the range D is more than L, to the total number M_0 of the measuring points. Figure 6.19c shows the probability $P(D > L)$ (in percentage scale) as a function of the range L (in meters) calculated from the measurement results for the vertical polarization of the transmitting wave (horizontal plane). Similar plots for horizontal polarization are shown in Fig. 6.19d. For example, the probability that the maximum range would be more than 100 m for antenna 1 $P(D > 100)$, as observed in Fig. 6.19a and b is 87 and 72 % for horizontal and vertical polarizations, respectively, while for the diversity technique, the corresponding values are equal to 98 and 90 %.

6.5 MIMO Capacity Estimation for the Vehicle-Mounted Antenna Systems

6.5.1 Distributed Roof-Mounted Four-Element Array

This section describes the MIMO performances for two distributed antenna configurations operating at 5.9 GHz band and mounted on the car [33, 34] that are proposed for MIMO automotive system. The array consists of three antennas mounted on the car in different positions. For the first configuration, a monopole antenna is placed at the center of the roof at a height of 1.4 m above the ground, whereas the square patch antenna elements are mounted to the front and left side mirrors, as shown in Fig. 6.20a. The patch antennas are vertically polarized with

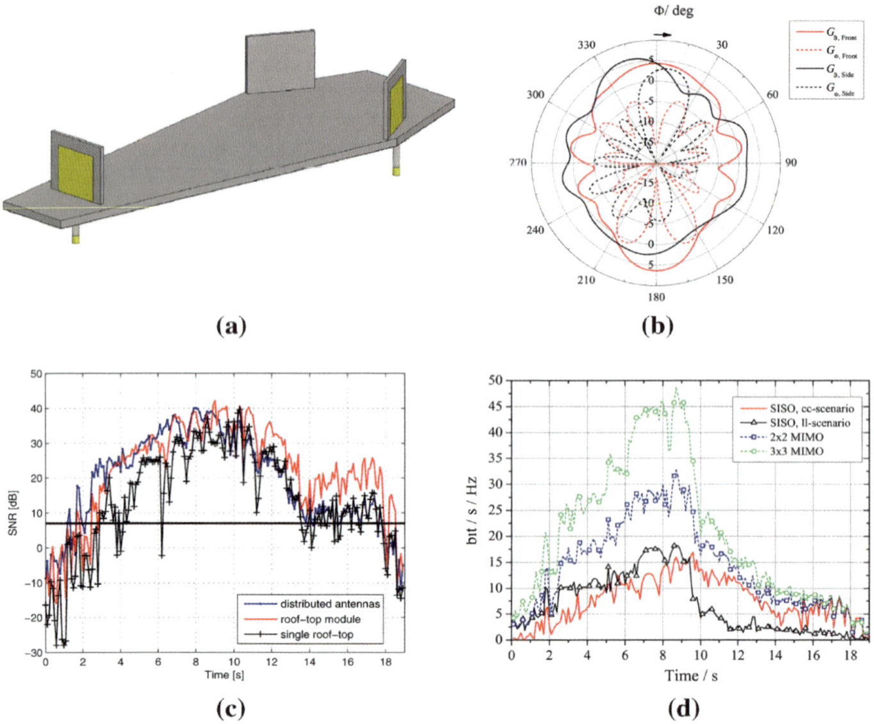

Fig. 6.21 **a** Distributed roof array topology with three patch elements; **b** Antenna element gain in dBi in horizontal plane for $\theta = 70°$; **c** SNR curves for maximum ratio combining diversity, **d** MIMO spectral efficiency for a few communication scenarios, [33, 34] © 2010 IEEE

the main beams oriented in the direction of the motion. The distance between the side mirror antennas and the roof antenna is 137 cm; the distance between side mirrors is 167 cm. The gain radiation patterns for both components G_θ and G_ϕ, as the functions of the azimuth angle ϕ for θ=90°, are shown in Fig. 6.20b. The driving direction X corresponds to the angles $\phi = 0$, $\theta = 90°$. The gain radiation patterns for both components G_θ and G_ϕ as the functions of the elevation angle θ for azimuth angle $\phi = 0°$ are shown in Fig. 6.20c. The patch antenna [35] provides the θ gain component value of about 5 dBi for the elevation angles θ between 40 and 140°, whereas the monopole gain value for the θ component is -1 dBi in the horizontal plane. In contrast to an ideal monopole, the gain maximum is shifted to higher elevation angles and additional sidelobes appear due to the finite size of the ground plane.

A significant ϕ gain component value is caused by the asymmetrical ground plane (the monopole antenna is mounted on the roof close to the rear window, as shown in Fig. 6.20a). Common topology of the patch antenna element is shown in Fig. 6.20d.

The second antenna array geometry includes three identical patch antennas shown in Fig. 6.21a. A module with elements is positioned at the same place where the first configured monopole antenna is placed. The distance between the side antennas and the front antenna is 9.2 cm, and the distance between the side antennas is 4 cm. Square patch antennas are optimized to maintain impedance matching within DSRC and WLAN applications. The horizontal gain patterns of the front and rear left antenna elements for both polarization components G_ϕ and G_θ presented in Fig. 6.21b are drown for maximum antenna gain for elevation angle $\theta = 70°$.

For an engineer, who designs MIMO antenna array system for vehicle applications, it is desirable to estimate (using simulation methods or experimentally) the maximum capacity value that characterizes the data transmission rate. The use of the MIMO system can be beneficial when the propagation channel between the transmitting and receiving antennas suffer from multipath reflections, as seen in the following study. Radio propagation effects and antenna influence on the channel are based on the developed software [36], which models two communicating vehicles under LOS conditions. It is assumed that they share information about the traffic situation and move in opposite directions on the same major road. The operating frequency is 5.9 GHz and the transmitting power is 23 dBm. Detailed parameters for the simulation scenario are presented in the papers published by Communication Workshop (ICC), 2010 IEEE International Conference [37]. Figure 6.21c [33] shows SNR for maximum ratio combining diversity (MRS) method when using the following communication scenario. At the beginning of the simulation, the vehicles move along different perpendicular major roads toward the intersection under nonline of sight (NLOS) conditions. Since one vehicle (RX) receives a warning message, it breaks down to yield right of way to the other vehicle (TX). The vehicles pass to each other at the point corresponding 9.5 s. After TX turned left at the intersection, the vehicles accelerate and move in opposite directions. The curve marked as distributed antennas is obtained for the array shown in Fig. 6.20a, a rooftop module plot is related to the mounting demonstrated in Fig. 6.21a, and a single rooftop graph is obtained for single rooftop monopole antenna. It is clearly seen that both array topologies lead to higher SNR in comparison with the single rooftop antenna. Figure 6.21d [34] demonstrates spectral efficiency (capacity normalized to the bandwidth), which calculated according to the formulas presented in Chap. 4, for three systems: SISO, 2×2, 3×3 MIMO. SISO cc scenario uses roof isotropic monopole antennas (from transmit and receive sides) mounted at the origin of the coordinate system (Fig. 6.20a), SISO ll scenario graph is obtained for transmitting and receiving patch antennas located in position 3 (Fig. 6.20a), 2×2 scenario uses one roof and left side patch antennas, and 3×3 system uses all tree antennas. As expected, 3×3 system shows significant (3 times) increase in spectral efficiency in comparison with the reference topology. However, this benefit rapidly drops after the cars have passed each other, since the information contribution added by the two patch antennas significantly decreases.

6.5.2 Comparison of SISO, SIMO, and MIMO Systems for Moving Cars

This section presents the results of a capacity evaluation for a MIMO antenna system that realizes the communication between moving cars in urban environment [37–44]. The evaluation is based on a simulation of radio a propagation channel which takes into consideration the multipath propagation and dynamic road traffic model [38]. The simulator has three main parts: a road traffic model, comprehensive model of the road environment, and accurate model for multipath wave propagation between the transmitting and receiving vehicles. Traffic model provides realistic description of the traffic and considers the motion of every vehicle with different velocities and acceleration and deceleration. Ray-optics method is used for channel description [39]. Ray-optics model is based on the assumption that the wavelength is small compared to the dimensions of the model objects in the simulation scenario. The Fresnel reflection coefficients are used to model the reflections. To describe scattering from the trees the surface of the tree model is subdivided into small squared tiles.

Vertically polarized omnidirectional antennas have been assumed to operate at 5.9 GHz. The simulation scenario [37] consists of four lanes of traffic and vicinity of parked cars, houses, traffic signs, and trees. The transmitter is located on one of the middle lanes and wave propagation is calculated for evenly distributed receiving points along the road. All points on the four lanes are considered as potential receivers. The distance between the transmitter and the receiver is <200 m. The spacing between the receiving points is chosen equal to 5λ to provide good interpolation when modeling the wave propagation between those points. The receiving points have been investigated for five different heights: 30, 60, 90, 120, and 150 cm. Receiver and transmitter are always at the same heights. To evaluate the performances of the single and multiple antenna systems, the channel capacity was calculated. The capacity is a function of SNR and the channel matrix H, which describes transfer coefficients between the transmitting and receiving antenna elements. White noise power calculated for 10 MHz bandwidth and room temperature 20° C is equal to -103 dBm. The transmitted power for single transmit antenna is equal to 23 dBm. Figure 6.22 shows the simulation results: Fig. 6.22a demonstrates results for the SISO system, Fig. 6.22b shows the calculations for the SIMO system, and Fig. 6.22c presents results for the MIMO 2×2 technique.

To calculate the capacity for SISO system, the well-known Shannon formula has been used.

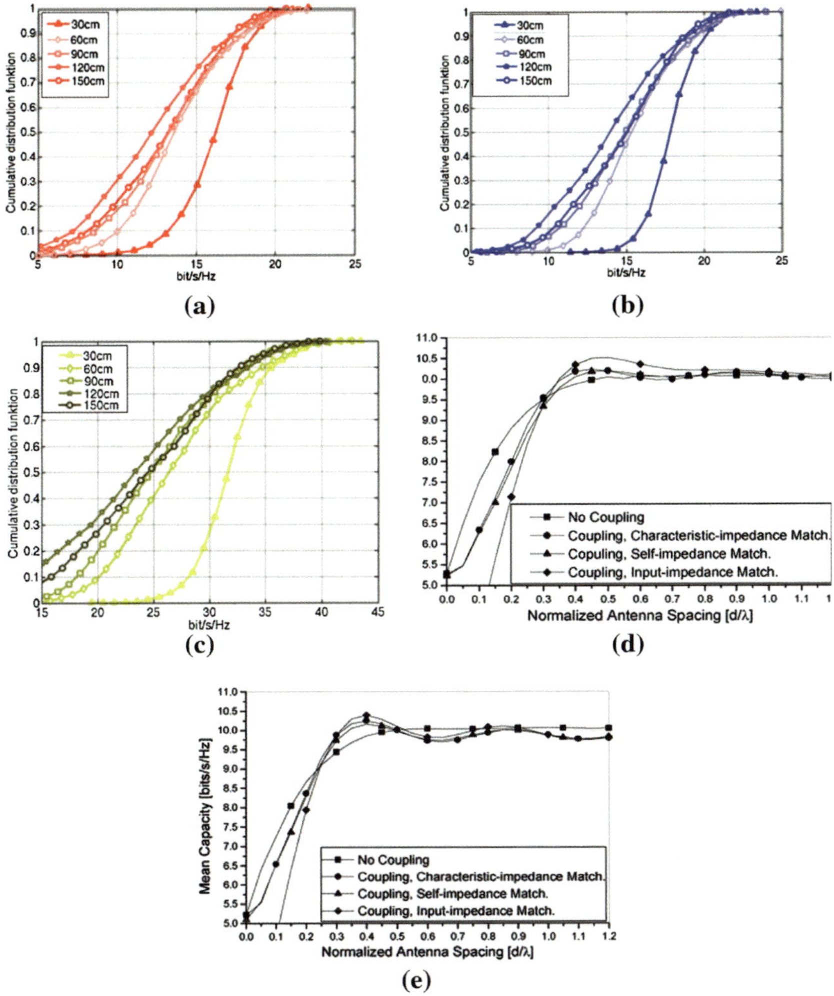

Fig. 6.22 **a** Results for SISO system; **b** SIMO design; **c** 2 × 2 MIMO topology; **d** Mean capacity versus antenna spacing with mutual coupling effect in urban street area; **e** Mean capacity at the Interstate Highway area, [37, 46] © 2009 IEEE

$$C = \log_2\left(1 + SNR \cdot |h|^2\right) \tag{6.1}$$

SIMO design computer simulations are obtained for maximum ratio combining which is described in Chap. 3.

$$C = \log_2[\det(\mathbf{I} + SNR \cdot \mathbf{H} \cdot \mathbf{H}^*)] \tag{6.2}$$

For the 2 × 2 MIMO system, the transmitting power for each antenna is halved compared with the SISO system

$$C = \log_2\left[\det\left(\mathbf{I} + \frac{SNR}{2} \cdot \mathbf{H} \cdot \mathbf{H}^*\right)\right] \tag{6.3}$$

Figure 6.22 shows the cumulative distribution function, which defines the probability that a real-valued random capacity variable will be found at a value less than or equal to a certain value.

From the simulations it can conclude that the 30 and 150 cm antenna heights show better results in comparison with 60, 90, and 120 cm. As it follows from the reference paper [40], this result is explained by better LOS propagation conditions for the heights 30 and 150 cm. Also, it is clearly seen from the results that 2×2 MIMO system is significantly better than SISO or SIMO design.

It is necessary to emphasize that this simulation reflects only a rough approximation of the real-world environment. However, it gives an idea of the benefits of utilizing the MIMO system in C2C (V2V) communication.

6.5.3 *Impact of Mutual Coupling on MIMO Vehicle-to-Vehicle System*

As we noted earlier, MIMO capacity is determined by the propagation channel and antenna array parameters. In V2V communication, robust compact antennas with reduced visual presence and aerodynamic drag can be utilized as the components of the MIMO system. Therefore, transmitting and receiving arrays with small interelement spacing are desired. Zajic [45] from the US Naval Research Laboratory presented interesting investigation that analyzes the 2×2 MIMO system in the vehicle environment with variable spacing between antenna elements. The analysis was confirmed by experimental results on urban surface streets (USS) and Interstate Highway (IH) road surfaces [45–48]. It is assumed that both the transmitters and receivers are in motion and equipped with vertically polarized half-wavelength electric dipole antenna elements. The MIMO system is operating at the WLAN frequency band equal to 2.45 GHz ($\lambda = 123.3$ mm). Figure 6.22d and e show the mean capacity value versus antenna spacing for USS and IH environments, respectively. The instantaneous capacity is averaged over 5,000 channel realizations. Plots are drawn for a few different conditions: (a) no coupling between the elements; (b) antenna element is matched to the characteristic impedance of transmission line load Z_0 (characteristic match); (c) antenna is matched to the complex conjugate match value, but without mutual coupling (self-impedance match); (d) input impedance is matched with complex conjugate value taking into consideration coupling effect (input impedance match). It is seen that the capacity in USS and IH environments are relatively similar, and for the element spacing more than 0.3λ, the system with mutual coupling does not significantly differ from the system with decoupling antenna elements. Also, it is seen from the figures that different matching techniques have no significant effect on the capacity values, even for the no coupling case. Again, we would like to emphasize that these results take into consideration propagation channel parameters in a vehicle environment and the mutual coupling effect between antenna elements.

References

1. Sugiura S, Iizuka H (2007) Reactively steered ring antenna array for automotive application. IEEE Trans Antenna Propag 55(7):1902–1908
2. Sawaya K (2003) Review of research and development on linear antennas, IEICE transaction communication. V E86-B 3:892–899
3. Nakano H et al (2006) A small steerable beam antenna. International symposium on antennas and propagation ISAP
4. Ohira M et al. (2011) Low-profile antenna structure, US Patent # 7,956,815, Publication Date June 2011
5. Zhang Z et al. (2008) An endfire phased array used in wireless access for vehicular environments (WAVE). Microwave and millimeter wave technology, 2008, ICMMT, international conference, pp 428–431
6. Zhang Z et al (2010) An endfire beam-switchable antenna array used in vehicular environment. IEEE Antennas Wirel Propag Lett 9:195–198
7. Gatsinzi M et al (2007) Study of a 5.8 GHz frequency band patch antenna integrated in a vehicle for automotive DSRC applications. ICEAA, Turin, Sept 2007
8. Gatsinzi M et al (2008) Inter-vehicle communications: study of the integration of a 5.8 GHz antenna and radio link in the 5 GHz Band. Antennas and propagation society international symposium
9. Huang Y (2009) A compact smart antenna for WiMAX radio. Mobile WiMAX symposium, 2009. IEEE, pp 169–173
10. Thiel A et al In Situ Vehicular Antenna Integration and Design Aspects for Vehicle to Vehicle Communications, http://publik.tuwien.ac.at/files/PubDat_193865.pdf
11. Thiel A et al. Initial results of multi-element antenna performance in 5.85 GHz vehicle to vehicle scenarios. In: proceedings of 38th european microwave conference, 2008, Amsterdam
12. Posadas V et al. Approximate analysis of short circuited ring patch antenna working at TM_{01} mode. IEEE Trans Antennas Propag 24(6):1875–1879
13. Deo P et al. (2009) Beam switched loop antennas-square to circular. Microwave conference 2009, pp 1871–1874
14. Raaza A et al. (2008) A novel 8 feed beam switched antenna. Antennas and propagation society international symposium, IEEE, 2008
15. McEwan N et al (2003) A new design of horizontally polarized and dual-polarized uniplanar conical beam antennas for HIPERLAN. IEEE Antennas Propag 51(2):229–237
16. Negut A et al. (2009) Performance of a 20 cm short active AM/FM monopole antenna for automotive application. Antennas and propagation (EUCAP) 2009 3th European conference, 2009, pp 2708–2712
17. Negut A et al. (2010) High-impedance amplifier for a novel 14 cm short AM/FM automotive active antenna. Antennas and propagation (EUCAP) 2010 4th European conference, 2010, pp 1–5
18. Kittinger G et al (2011) Antenna diversity system, US Patent Application No. 2011/0187613, Aug 2011
19. Sliskovic M et al (2010) Roof antenna array for phase diversity system, US Patent No. 7,696,939, Publication Date April 2010
20. Brzeska M et al. (2009) New generation of in-mirror integrated antennas. Antennas and propagation (EUCAP) 2009 3th European conference, 2009, pp 2704–2707

21. Itoh N, Tsuchida K (2006) HDTV mobile reception in automobiles. Proc IEEE 94(1): 274–280
22. Compact Antenna with Novel High Impedance Amplifier-Diversity Module for Common Integration into Narrow Dielectric Parts of a Car Skin. Institute of High Frequency Technology and Mobile Communication University of the Bundeswehr Munich, Germany
23. Iizuka H et al. (2005) Modified H-shaped antenna for automotive digital terrestrial reception. IEEE Trans Antennas Propag 53(8, Part 1):2542–2548
24. Kudo K et al. (1999) TV antenna apparatus for vehicle, US Patent 5, 977,919, 1999
25. Wang J, Winters J (2004) An embedded antenna for mobile DBS. Vehicular technology conference, 2004, IEEE, 60th, pp 4092–4095
26. Hules F, Streelman G, Yen H (2005) A direct broadcast reception system for automotive OEMs. Antennas and propagation society international symposium, IEEE pp 80–83
27. Wang J et al. (2007) Vehicle mounted satellite antenna embedded within moon roof or sunroof, US Patent No. 7,227,508, June 2007
28. Wang J et al. (2008) Vehicle mounted satellite antenna embedded within moon roof or sunroof, US Patent No. 7,391,381, June 2008
29. Rio D et al. (2010) Sub-array polarization control using rotated dual polarized radiating elements, US Patent Application No. 2010/0253585, Oct 2010
30. Tiezzi F et al. (2011) Method and apparatus for a compact modular phased array element, US Patent Application No. 2011/0025574, Feb 2011
31. Vaccaro S et al. (2011) Low cost ku-band electronic steerable array antenna for mobile satellite communications. Antennas and propagation (EUCAP) proceedings of the 5th European conference, 2011, pp 2362–2366
32. Rabinovich V et al. (2009) Antenna system for remote control automotive applications, US Patent 7,564,415, Publication date July 2009
33. Scack M et al. (2010) Analysis of channel parameters for different antenna configurations in vehicular environments. Vehicular technology conference, IEEE
34. Kornek D et al. (2010) Effects of antenna characteristics and placements on a vehicle to vehicle channel scenario. ICC 2010 workshop on vehicular connectivity, CapeTown, South Africa, May 2010
35. Amman M (1999) Square planar monopole antenna. IEE national conference on antennas and propagation, pp 37–40
36. Schumacher H et al. (2009) Coupling of simulators for the investigation of car to X communication aspects. Second IEEE international workshop on vehicular networking (VON-09), Dec 2009
37. Reichardt L et al. (2009) Performance evaluation of SISO, SIMO and MIMO antenna system for car-to-car communications in Urban environments. Intelligent transport systems telecommunications 2009, 9th international conference, pp 51–56
38. Reichardt L et al. (2010) Simulation and evaluation of car-to-car communication channels in urban intersection scenarios. Vehicular technology conference, 2010 IEEE
39. Wiedemann R (1974) Simulation des Strassenverkehrsflusses, (in German). Tech Report, 1974
40. Fügen T et al (2006) Capability of 3D Ray-tracing for defining parameter sets for the specification of future mobile communications systems". IEEE Trans Antennas Propag, part 1 54(11):3125–3137
41. Matthaiou M et al. (2008) Capacity study of vehicle-to-roadside MIMO channels with a line-of-sight component. IEEE wireless communications and networking conference, 2008, pp 775–779
42. Laurenson D et al. (2009) Modelling for vehicle to vehicle applications. Antennas & propagation conference, 2009. LAPC 2009. Loughborough, pp 42–48
43. Sarris I, Nix A (2007) Design and performance assessment of high-capacity MIMO architectures in the presence of a line-of-sight component. Vehicular Technology. IEEE Trans 56(4):2194–2202

44. Sarris I, Nix A (2006) Design and performance assessment of maximum capacity MIMO architectures in line-of-sight. Communications, IEE Proc 153(4) pp 482–488
45. Zajic A (2009) Effects of mutual coupling on capacity of MIMO vehicle-to-vehicle systems. Military communications conference, MILCOM 2009, IEEE
46. Zajic A (2009) Impact of mutual coupling on capacity of MIMO vehicle-to-vehicle systems. Global telecommunication conference, 2009, IEEE
47. Zajic A (2008) Statistical modeling and experimental verification for wideband MIMO mobile-to-mobile channels in urban environments. Telecommunications, 2008. ICT 2008. International conference on, 2009
48. Zajic A (2008) Statistical modeling and experimental verification of wideband MIMO mobile-to-mobile channels in highway environments. Personal, indoor and mobile radio communications, 2008. PIMRC 2008. IEEE 19th international symposium on, 2008

Chapter 7
Radar Arrays for Vehicle Applications

7.1 Introduction

Rapid progress in radar technology has a direct impact on the design issues of high performance systems. Radar performs well in all weather conditions and, therefore, is utilized in a variety of applications: short- and long-range automotive radars (SRR and LRR), traffic monitoring, the automotive radars for intelligent cruise control [1], etc. Some of these applications are demonstrated in Fig. 1.3a.

For example, 24 GHz SRR system with range operation below 30 m is used to provide side, rear, front, blind spot, and other types of safety enhancements. Long-range radar (LRR) operates at 77 GHz, provides the range up to 150 m and can be used for adaptive cruise control (ACC) applications.

Traffic monitoring systems, which are used to detect vehicle speed, can operate with an unmodulated (CW) signal. The receiver in such a circuit processes the signal reflected from the vehicle. Due to the Doppler Effect, the mutual speed of the car related to the antenna causes the frequency shift that can be determined by

$$\delta f = 2 \cdot v \cdot \frac{f}{c} \cdot \cos \alpha \tag{7.1}$$

where v is the speed of the car, f the operating frequency, c the light speed, and α the angle between the speed vector and the vector that determines the direction along the line between the vehicle and the radar. When the vehicle and radar are getting farther apart, the observed frequency is lower than the transmit frequency ($\alpha > 90°$) and when the vehicle and transmitter are getting closer together, the observed frequency is higher than the transmitting frequency ($\alpha < 90°$). Such a Doppler system is the heart of the well-known police speed radars with a typical range up to 100 m, when using a transmitter power of 10–100 mW.

Frequency-modulated continuous wave pulsed radars are arranged in monostatic or bistatic modes and provide information on speed and distance. The monostatic arrangement utilizes a single antenna to transmit and receive, while the bistatic system has two separate antennas, one connected to the transmitter and

V. Rabinovich and N. Alexandrov, *Antenna Arrays and Automotive Applications*,
DOI: 10.1007/978-1-4614-1074-4_7,

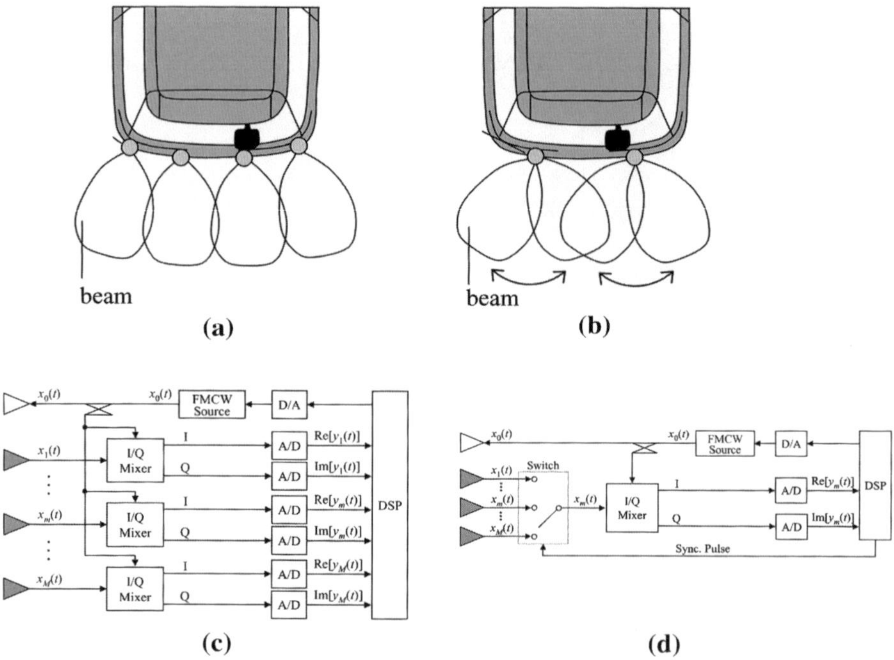

Fig. 7.1 **a** Fixed beam array concept. **b** Electronically scanned beams. **c** Simplified block diagram of the digital radar with multi-receiver configuration. **d** Simplified schematic of the radar with single receiver

the other to the receiver. The angle information can be obtained using a directional receiving antenna array. Directional arrays can be arranged as fixed beam systems or electronically scanned arrays as shown in Fig. 7.1a and b. An electronically scanned system can be designed based on a regular phased array, multiple-channel array with digital beam processing, as shown in Fig. 7.1c, or by using the simplified switched topology presented in Fig. 7.1d.

The radar cross section (effective reflected area) of a car is an important factor for radar-range estimation because it determines the maximum detectable range of the automotive radar. It is well known that the received radar power signal reflected from the object in free space is calculated by the following expression:

$$P_r = \frac{P_t \cdot G_t}{4 \cdot \pi \cdot R^2} \cdot \sigma \cdot \frac{1}{4 \cdot \pi \cdot R^2} \cdot \frac{G_r \cdot \lambda^2}{4 \cdot \pi} \tag{7.2}$$

where P_t and G_t are power and the gain of the transmitting antenna toward the vehicle, respectively; P_r and G_r are received power and receiving antenna gain, R is the distance between the radar and a vehicle, λ the wavelength, and σ the vehicle radar cross section (RCS) value toward the radar. It is seen that the received power is proportional to the RCS of the car: a low cross section value leads to the low level of the received power and, as a result, reduced radar operating range.

Table 7.1 Detection of range values for different radar applications

Applications	Detection range (m)	Safety aspect	Technology
Adaptive cruise control	200	Normal driving; accident avoidance	77 GHz Radar
Pre-crash	30	Accident; mitigation of impact	77 GHz/24 GHz Radar; 76 GHz/81 GHz Radar
Blind spot detection	20	Normal driving; accident avoidance	24 GHz Radar/vision sensor
Lane departure Warning	60	Normal driving; accident avoidance	Vision sensor
Stop and go	30	Normal driving; accident avoidance	77 GHz/24 GHz Radar; 76 GHz/81 GHz Radar

Formula (7.2) is obtained assuming free space propagation. In a real ground environment, the signal strength at the receiver side can be reduced by ground reflections. Power from the transmitting antenna and the scattering object reaches the receiving antenna via ground reflection in addition to the direct path. The level of reflections depends on the angles of incidence. For example, for incidence angles between 5 and 0.5°, the reflection coefficients for horizontal and vertical polarizations are similar and about 0.7–0.8 in magnitude and the phase shift is close to 180° [2]. Let us replace formula (7.2) with the following ratio:

$$P_r^{\text{ground}} = P_r \cdot \beta \tag{7.3}$$

where factor β takes into account ground effect. Maximum radar range $R_{\max}$ can be made using expressions (7.2) and (7.3).

$$R_{\max} = \left[\frac{P_t \cdot G_t \cdot G_r \cdot \sigma \cdot \beta \cdot \lambda^2}{(4 \cdot \pi)^3 \cdot P_r^{\min}}\right]^{1/4} \tag{7.4}$$

As an example, estimate the maximum range for SRR assuming the following system parameters: transmitting power $P_t = 10$ dBm; transmitting and receiving antenna gains are equal $G_t = G_r = 25$ dBi; wavelength corresponding to the operating frequency 25 GHz is equal to 1.2 cm. We will make calculations for the worthy case when the direct signals and those reflected from the ground reach the object and the receiving antenna in opposite phases (180° phase shift) and we also assume that reflection coefficient from the ground is equal to 0.8. In this case, factor β is equal to $(0.2)^4 = 0.0016(-28$ dB). The number of rough measurements [3–5] show that the radar RCS of the road vehicle varies within the interval 10 m^2 to (1/10) m^2, depending on the aspect angle. We estimate (7.4) for the highest value of RCS equal to 10 m^2 that is obtained for the front (0°), rear (180°), and side (90° or 270°) aspect angles. Value $P_r^{\min}$ in formula (7.4) determines the minimum detectable signal, which depends on the receiver sensitivity. Assuming that $P_r^{\min} = -95$ dBm, we obtain that $R_{\max} \approx 50$ m. According to the requirements for

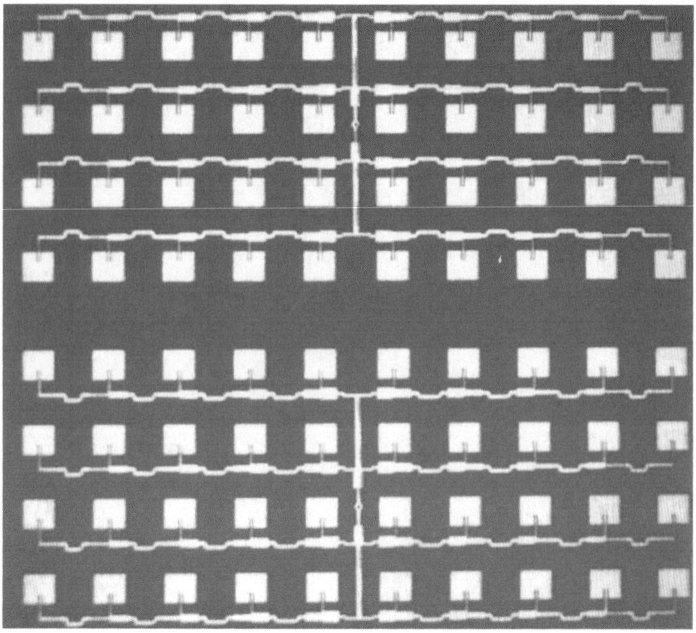

Fig. 7.2 Planar double antenna array system for 24 GHz, [7] © 2010 IEEE

SRR [6], the maximum range is about 20–50 m. Of course, this is a rough estimation; however, example can help an engineer to make a quick estimation of the radar range as a function of the main system parameters. Table 7.1 shows typical detection range values for various short- and long-range radars that are required for automotive safe applications.

A lane departure warning system (Table 7.1) designed to warn a driver when the vehicle begins to move out of its lane on freeways and arterial roads typically uses vision sensor instead of microwave system.

7.2 Brief Review of Anti-Collision Radar Automotive Antenna Arrays

7.2.1 Patch Double Planar Topology

Figure 7.2 shows an example of a planar double antenna system that consists of two identical patch arrays etched on PCB dielectric material.

One array operates as a transmitting antenna and the other functions as a receiving sensor. A parallel feeding network provides a matching design with minimal reflections (typical return loss less than 20 dB). A topology designed for

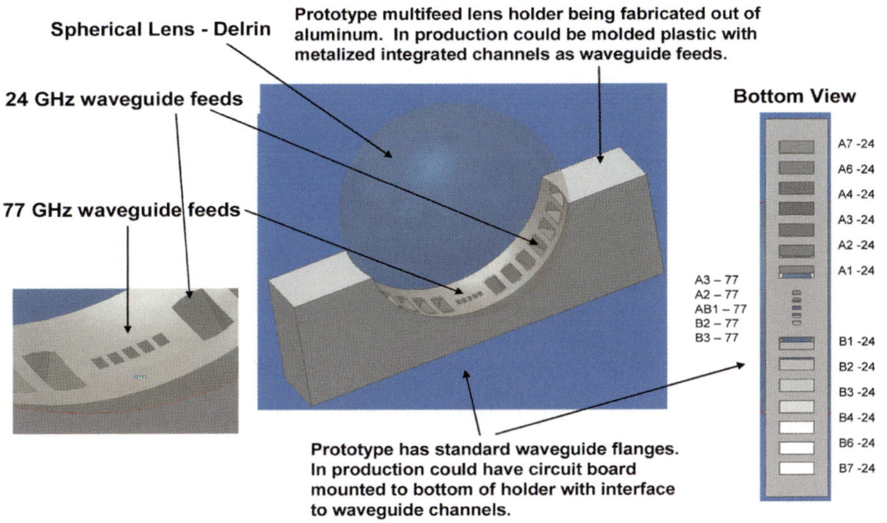

Fig. 7.3 Integration of LRR and SRR systems; [8] © 2007 IEEE

24 GHz with 14 × 6 patch elements [7] has the gain of 26 dBi, in the E-plane −3 dB beamwidth is 6° and in −3 dB H-plane beamwidth is equal to 18°. Inter-element space is of equal wavelength in free space.

7.2.2 Combined Long- and Short-Range Spherical Lens Configuration

Figure 7.3 demonstrates an array aperture developed for the integration of LRR and SRR systems [8].

It consists of a spherical dielectric lens antenna with multiple feeds located around it. The spherical lens with 3.62 cm diameter has a dielectric constant of approximately 3. The length is surrounded by waveguide feeds positioned to create the 10° outer beams at 24 GHz and 3.5° beams at 77 GHz. This multibeam antenna system has a directivity of about 26 dBi for 24 GHz band and 33 dBi for 77 GHz. The azimuth and elevation beamwidth for 24 GHz are around 10° and 3.3° for 77 GHz. The multibeam system covers azimuth angle sector equal to −75 to 75° for 24 GHz and −6 to 6° for 77 GHz.

7.2.3 Yagi-Uda Array for 76 GHz Band

Figure 7.4a illustrates a car mounted planar end-fire Yagi-Uda antenna array with RF integrated circuits for operation within the 76–77 GHz band [9, 10].

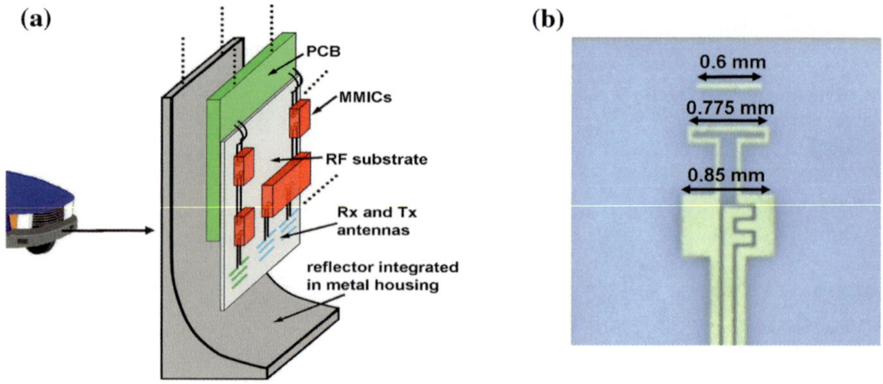

Fig. 7.4 **a** Radar system concept with antenna array and cylindrical parabolic reflector. **b** Yagi-Uda antenna element; [9] © 2010 IEEE

The array is printed on high-permittivity and low loss alumina substrate with a dielectric constant of about 10. Each of the three identical planar array elements is shown in Fig. 7.4b. The measured gain of a single antenna element is about 5.4 dBi, the HPWB is 65° in the E-plane, and 54° in the H-plane. The system is mounted on a car with a cylindrical parabolic reflector, which focuses the beam in the elevation plane (H-plane), and a beam focusing in the azimuth plane is provided by the array in the E-plane.

The measured element HPWB with the reflector becomes 7° in the H-plane. With three array elements the measured HPWB in the E-plane is 30°.

7.2.4 Parabolic Transreflector Array System

Figure 7.5a demonstrates a schematic diagram of the low profile 76–77 GHz antenna array module with a parabolic transreflector for beam collimation [11]. The module is mounted on the reverse side of the antenna ground plane with a few (three in the design presented in the paper [11]) waveguide feeds.

A system with 145 mm diameter has a resolution of about 3.2 m at a distance of 100 m. The waveguide ports of the module radiate a millimeter wave toward the transreflector on the inner surface of the lens with the polarization orthogonal to the polarization of the reflector. This radiation is reflected by the transreflector toward the twist-reflector (named as twist-circuit in the figure), which in turn reflects it back to the transreflector with its polarization rotated at 90° so that it passes through the reflector. The parabolic lens collimates and focuses energy before it is radiated from the antenna. A detailed description of the transreflector and twist-circuit design is presented in patent [12].

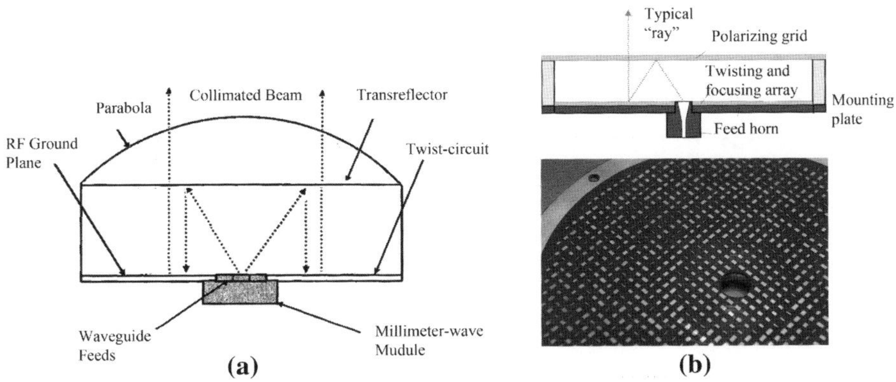

Fig. 7.5 **a** Antenna array module with parabolic transreflector for beam collimation **b** Millimeter wave microstrip reflect array

7.2.5 Topology with Microstrip Reflect Array

The antenna array concept presented in Fig. 7.5b [13] is a compact design based on the millimeter wave microstrip reflect array. The key feature of this design is how the individual elements of the twisting and focusing reflect arrays are made to scatter with the desired phases. The basic design principle requires that the phase of the field reflected from an element in the reflect array be chosen so that the total phase delay from the feed to a fixed aperture plane in front of the reflect array is equal $2 \cdot \pi \cdot n$, where n is any integer number. Additionally, the wave reflected from the microstrip elements has to pass through the polarizing grid with minimal loss. The reflect array itself is realized on a 0.254 mm thick substrate with a dielectric constant of 2.2; unit cell size is 2.2 mm × 2.2 mm (see solid lines in Fig. 7.3). The beamwidths are about of 3° in both planes, maximum gain at this frequency is 33.7 dB, 3 dB bandwidth is 10 GHz, and the aperture efficiency 46 %.The diameter of the array operating at 76.5 GHz is 90 mm and the height is equal to 26 mm. Detailed calculation of the reflected wave phase, as a function of the patch size based on a full-wave moment method solution, was presented by Pozar [14] where experimental results for the operating frequency 28 GHz are also demonstrated.

7.2.6 Array Concept with Cylindrical Lens Configuration

Figure 7.6a shows the array concept using a cylindrical lens configuration. A cylindrical lens is illuminated by an E-plane oriented column of series fed microstrip patches, presented in Fig. 7.6b [15–18].

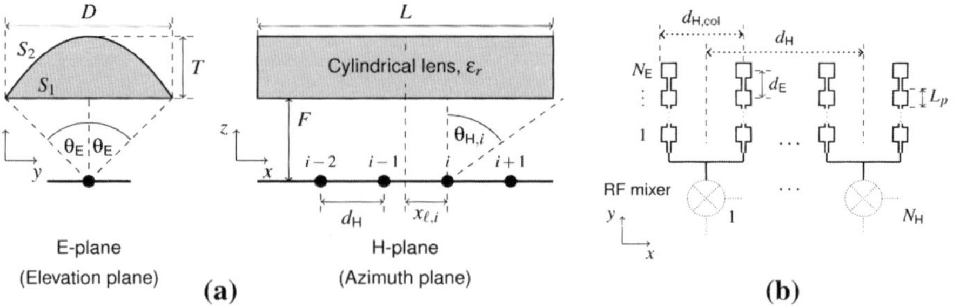

Fig. 7.6 **a** Cylindric dielectric lens antenna. **b** Subarrays of patch columns connected to individual RF mixers; [15] © 2008 IEEE

In the H-plane, receiving antennas form the uniform linear array of N_H patch subarray columns that are positioned on the focus line of the cylindrical lens with equidistant spacing d_H. This lens significantly simplifies the array design, reducing the number of radiators in E-plane N_E. A tapered aperture distribution in the E-plane for a low sidelobe level could be achieved with a tapered patch width, as shown in [16]. The column spacing is usually equal to half free space wavelength, which corresponds to 3.9 mm for operating frequency 77 GHz. Simulations [17] are conducted for −1.5 dB half power beamwidth in the E-plane (4.5°) and result in the following lens parameters for $N_E = 2$: $F = 15$ mm, $D = 40$ mm at the lens dielectric constant 3. The antenna gain is about 17 dBi and −3 dB beamwidth in the H-plane is about 30°. The experimental prototype of the array with 16 columns in the H-plane and dielectric cylindrical lens is described in reference paper [18]. The dimension of the lens in the azimuth plane is 94 mm. The antenna system has digital outputs and uses high resolution MUSIC and ESPIRIT algorithms for DOA estimation. Results show [18] that DOA estimation accuracy is less than 0.2° and the target range is up to 200 m for the radar cross sections of the objects of 10 m^2 with relative target speeds between −60 m/s and +20 m/s.

7.2.7 System with Rotman Lens

Figure 7.7a shows the sketch of the planar antenna array with the planar Rotman lens [19].

This system consists of a parallel plate region (body), antenna element ports connected with the array elements on one side, beam ports on the other side, and dummy ports on the side walls, as shown in Fig. 7.7a. The design of the lens is governed by the Rotman–Turner design equations, which are based on the geometry of the lens. The equations generate the positions of the antenna ports, beam ports, and dummy ports that are loaded with matched loads or with loss

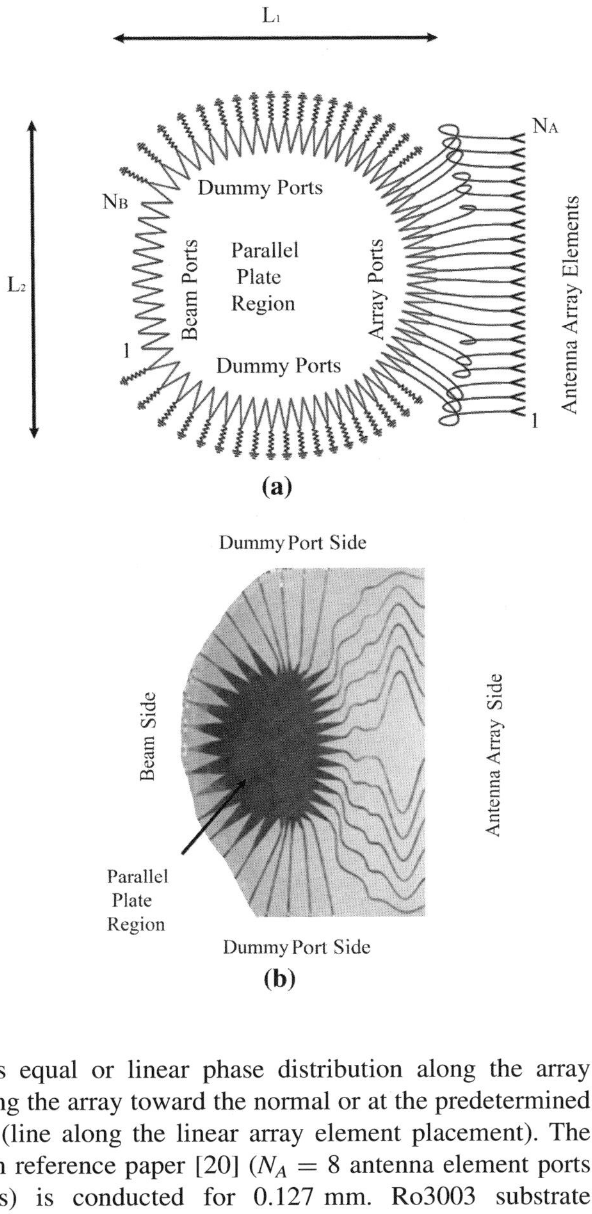

Fig. 7.7 Rotman lens: **a** Principle sketch. **b** Example of the etched Rotman lens body

material. The lens provides equal or linear phase distribution along the array elements, and allows focusing the array toward the normal or at the predetermined angle to the array baseline (line along the linear array element placement). The design example described in reference paper [20] ($N_A = 8$ antenna element ports and $N_B = 4$ beam outputs) is conducted for 0.127 mm. Ro3003 substrate (dielectric constant 3) and 50 ohm delay lines, parallel plate region size is $L_1 = 10$ mm by $L_2 = 8$ mm, frequency band 77 GHz. The resulting beam pointing angles are equal to $\pm 18°$ in azimuth. The loss of the lens with delay line network amounts to approximately 4 dB. An example of the Rotman lens body etched on the printed circuit board is presented in Fig. 7.7b.

7.3 Systems for Road Traffic Flow Estimation

Real-time road traffic density, especially in megacities is an important factor for signal control and effective traffic management. For example, congestion is a major problem for modern busy roads and highways. But building new roads to relieve that congestion is very expensive. Vehicle detectors are an integral part of the ITS [21]. Various detector technologies can currently provide traffic parameters, such as the presence and number of vehicles, as well as vehicle speed, for ITS systems. Early technologies are based predominantly on inductive loop detectors installed in the roadway subsurface [22]. A disadvantage of such detectors is their location in the roadway surface, where heavy traffic and construction can cause damage. These detectors are also inflexible; repair work is expensive, and requires closing traffic lanes. Overhead mounted video camera traffic monitoring systems, on the other hand, have been shown to provide several benefits [23, 24]. One benefit, such as, reduced maintenance costs, since maintenance can be performed with minimal traffic blockage. However, the algorithms for vision-based vehicle detection are not perfect when the weather or light conditions are not ideal [25–28].

A great advantage of microwave radar is that it is generally insensitive to inclement weather. The basic premise of this technology is that the microwave transmitter sends a signal toward the roadway. When a vehicle passes through this electromagnetic field, some of the signal energy is reflected back to the receiving antenna, providing a momentary pulse that indicates a vehicle has been detected. On the other hand, microwave radar applications must ensure that the receiving antenna beamwidth and transmitted waveform are suitable for the application. For example, only narrow beamwidth radar can monitor a single lane of flowing traffic. Some microwave detectors use a Doppler-shift principle. In this case, the receiver senses a change in signal frequency, thus, detecting both the vehicle and its speed. For any radar planned for traffic flow estimation, the key issue is analyzing the changes of the echo power amplitude and/or power spectrum between the vehicles and the backgrounds. Another microwave technology uses a remote traffic microwave sensor (RTMS) with a frequency-modulated continuous wave beam. RTMS operates as a presence detector to detect motionless vehicles, but it can also be used to detect vehicle speed. For example, Wavetronix developed Smart radar sensor (Model 125). which can detect up to 10 lanes of traffic and has a detection range of 250 feet. It also has the ability to measure traffic volume and individual vehicle speed using two receive antennas positioned side by side with enough space between them to create two separate, high definition beams. By measuring the time it takes for a vehicle to pass between the two antennas to within a fraction of a millisecond, Smart Sensor HD provides a highly accurate measurement that is then used to calculate each individual vehicle's speed. Simulations for smart radar with high resolution digital antenna array are presented in the reference paper [29]. The radar operates at a frequency band of 25 GHz ($\lambda = 1.2$ cm), includes nine element linear array with half-wave inter-element spacing, linear FM signal with sweeping bandwidth 100 MGz, and pulse period 0.75 ms. The sample frequency is

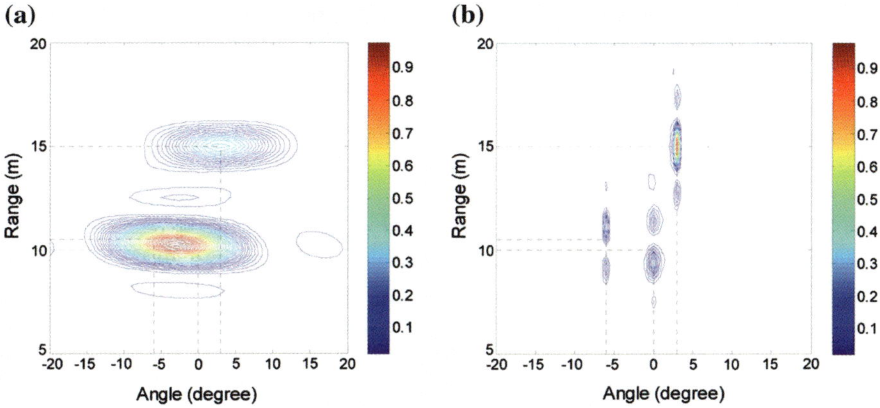

Fig. 7.8 **a** Imaging via conventional beamforming. **b** Imaging via high resolution algorithm; [28] © 2010 IEEE

Table 7.2 Vehicle detection results

Cars in lane	Cars missing	False alarm	Out-lane detected	Totally detected
417	5	0	45	467

500 kHz, which satisfies Nyquist sample rate. The image observations include three targets at the following angles: $\theta_1 = -6°$, $R_1 = 10.5$ m; $\theta_2 = 0°$, $R_2 = 10$ m; $\theta_3 = 3°$, $R_3 = 15$ m. Figure 7.8a is an imaging solution using conventional beamforming. The dotted line illustrates the true position of the targets. Conventional Rayleigh resolution is determined by the beamwidth of the antenna

$$\Delta\theta \approx \lambda / L_0 \tag{7.5}$$

where L_0 is the receiving antenna aperture and $\lambda =$ wave length. Because angle difference and range difference between targets 1 and 2 are all smaller than their corresponding resolutions, limited by antenna aperture and sweep bandwidth, these two targets are not correctly resolved. Figure 7.8b shows the results of high resolution method, as described in Chap. 3. As we can see, all three targets are correctly resolved with the high resolution method versus the conventional method where targets located at distances 10.5 and 10 m are not resolved.

7.4 Experiment Results of High Resolution System for Traffic Flow Estimation

The pulse radar with digital output processing for traffic flow estimation [25] operates at the carrier frequency of 24.125 GHz, pulse width of 350 ps, sweep time of 20 ms, pulse repetition time of 250 ns, and peak power equal to 5 dBm. The radar has directional transmitting and receiving arrays, each consisting of six elements. The beamwidth along the traffic flow is about 13°, which enables to measure objects that are just below the radar, which is fixed above a single lane with a perpendicular distance of about 6 m to the lane. The existence of the passing vehicle is based on the number of reflecting echo points during a few sweeps. For example, a car moving with a speed of 100 km per hour along a highway passes 0.5 m during one sweep time. Hence, for a typical car length of 5 m the number of reflecting points with increased intensity from the vehicle is about 10. In the experiments, it was assumed that if, there are more than three reflecting responses stronger than a certain threshold level in 20 sweeps, the vehicle is present below the radar (one reflecting response corresponds to one sweep time). The results of the vehicle detection are shown in Table 7.2.

Data demonstrated in the table clearly show that the proposed technique resolves signals reflected from the metal body against the reflections from the ground.

References

1. Ozkurt C, Camci F (2009) Automatic traffic density estimation and vehicle classification for traffic surveillance systems using neural networks. Math Comput Appl 14(3):187–196
2. Shefer J et al (1974) Clutter-free radar for cars. Wireless World, pp 117–122
3. Palubinskas G et al (2004) Radar signatures of road vehicles. In: 2004 IEEE international geoscience and remote sensing symposium proceedings, pp 1498–1501
4. Palubinskas G, Runge H (2007) Radar signatures of a passenger car. Geosci Remote Sensing Lett, IEEE, pp 644–648
5. Palubinskas G et al (2005) Radar signatures of road vehicles: airborne SAR experiments, SAR image analysis, modeling and techniques. In: Francesko Posa X (ed.) Proceedings of SPIE, vol 5980
6. Gresham I et al (2004) Ultra-wide radar sensors for short-range vehicular applications. IEEE Trans Microw Theory Tech 52(9):2105–2121
7. Feng X et al (2010) K-band micro-strip antenna array applied in anti-collision radar. IEEE international conference, Communication Technology (ICCT), pp 1240–1243
8. Colburn J et al (2007) Multifunction aperture for vehicular radar integration. IEEE vehicular technology conference, pp 2042–2046
9. Beer S et al (2010) Planar Yagi-Uda antenna array for W-band automotive radar applications. Antennas and propagation society international symposium, IEEE
10. Beer S et al (2009) Novel antenna concept for compact millimeter-wave automotive radar sensors. IEEE Antennas Wirel Propag Lett 8:771–774
11. Huguenin G, Moore E (1997) Compact microwave and millimeter wave radar. US Patent 5,455,589, Publication 1995 and 5,680,139, Publication 1997

12. Greshman I et al (2001) A compact manufacturable 76–77 GHz radar module for commercial acc applications. IEEE Trans Microw Theory Tech 49(1):44–57
13. Menzel W et al (2002) Millimeter-wave folded reflector antennas with high gain, low loss, and low profile. IEEE Trans Antennas Propag 44(3):24029
14. Pozar D et al (1997) Design of millimeter wave microstrip reflectarrays. IEEE Trans Antennas Propag 45(2):287–296
15. Wenig P, Weigel R (2008) Analysis of a microstrip patch array fed cylindric lens antenna for 77 GHz automotive radar. Antennas and propagation society international symposium, IEEE
16. Freese J et al (2000) Synthesis of microstrip series-fed patch arrays for 77 GHz-sensor applications. Asia Pacific microwave conference, pp 29–33
17. Wenig P et al (2008) A dielectric lens antenna for digital beamforming and superresolution DOA estimation in 77 GHz automotive radar. International ITG workshop on smart antennas, pp 184–189
18. Richer M et al (2010) 77 GHz automotive beamforming radar with SiGe chipset. German microwave conference 2010, pp 210–213
19. Rotman W, Turner R (1963) Wide angle microwave lens for line source applications. IEEE Trans Antennas Propag 11:623–632
20. Schoebel J et al (2005) Design considerations and technology assessment of phased-array antenna systems with RF MEMS for automotive radar applications. IEEE Trans Microw Theory Tech 53(6):1968–1975
21. Klein LA (2001) Sensor technologies and data requirements for ITS. Artech House ITS Library, Norwood
22. Bullock D, Heymsfield E (December 1998) Innovative application of directional boring procedures for replacing inductive loop detectors. Autom Constr 8(2):143–148
23. Ingo RM (1989) Application of machine vision to traffic monitoring and control. IEEE Trans Veh Technol 38:112–122
24. Kastrinaki V, Zervakis M, Kalaitzakis K (2003) A survey of video processing techniques for traffic applications. Image Vis Comput 21:359–381
25. Huan Y et al (2005) A high-range- resolution microwave radar system for traffic flow rate measurement. In: Proceedings of the 8th international IEEE conference on intelligent transportation systems, pp 880–885
26. Zhang H et al (2008) A novel method for background suppression in millimeter-wave traffic radar sensor, In: Proceedings of the 8th international IEEE conference on intelligent transportation systems, pp 699–704
27. Zhang H et al (2008) Adaptive traffic lane detection based on normalized power accumulation. In: Proceedings of the 8th international IEEE conference on intelligent transportation systems, pp 968–973
28. Arnold D et al (2010) Systems and methods for monitoring speed. US Patent 742450, Publication Date 2008
29. Wang P et al (2010) FMCW radar imaging with multi-channel antenna array via sparce recovery technique. International conference, Electrical and Control Engineering 2010, pp 1018–1021
30. Lee M, Kim Y (2010) Design and Performance of a 24-GHz switch-antenna array FMCW Radar system for automotive applications. IEEE Trans Veh Technol 59(5):2290–2297

About the Lead Author

Victor Rabinovich was born in Moscow, Russia and received an MS degree in Electronic Engineering from the Moscow Institute of Physics and Technology (MIPT). In 1976, he graduated with a PhD specializing in Electromagnetic and Antennas. For over 20 years he was employed by a leading Moscow Scientific Industrial Corporation serving in the research and development department. There, he participated in a variety of projects, including the development of multi-element phased antenna arrays with electronically controlled beam and smart adaptive antenna array systems. Since 1995, he has been working in North America designing antennas for different automotive applications. He published more than 40 journal articles published in Russian and international journals, holds more than 30 patents; six new solutions are protected by US patents. As a lead author he has written the book "Automotive Antenna Design and Applications" which was published at 2010. His current research and design interests are automotive antennas including small antennas for multiband operations.

V. Rabinovich and N. Alexandrov, *Antenna Arrays and Automotive Applications*,
DOI: 10.1007/978-1-4614-1074-4,

Index

V. Rabinovich and N. Alexandrov, *Antenna Arrays and Automotive Applications*,
DOI: 10.1007/978-1-4614-1074-4,

V

Y

Printed by Printforce, the Netherlands